# THE LANGUAGE OF THE HORSE

MICHAEL SCHÄFER

# THE LANGUAGE OF THE HORSE
## Habits and Forms of Expression

*Translated by Daphne Machin Goodall*

KAYE & WARD · LONDON
ARCO PUBLISHING COMPANY INC · NEW YORK

photographs by the author

First published in West Germany 1974
First published in Great Britain by
Kaye & Ward Ltd
21 New Street, London EC2M 4NT
1975
First published in the USA by
Arco Publishing Company Inc.
219 Park Avenue South
New York, NY 10003, USA
1975

Copyright © 1974 Nymphenburger Verlagshandlung GmbH
English translation copyright © Kaye & Ward Ltd 1975

All rights reserved. No part of this publication may be reproduced, stored in a retrieval system, or transmitted, in any form or by any means, electronic, mechanical, photocopying, recording or otherwise, without the prior permission of the copyright owner.

ISBN 0 7182 1120 0 (Great Britain)
ISBN 0 668 03762 8 (USA)
Library of Congress Catalog Card Number 74-24802 (USA)

Printed in Great Britain by
Northumberland Press Limited, Gateshead

*For my wife,*

*without whom my books would not be written*

# CONTENTS

*Introduction*   1

## THE NATURAL HABITS OF THE HORSE

1. The Preoccupations of Equidae
   *The 'inner clock' of Equidae. Natural territory and size. The origin of habitat loyalty and the urge to roam. Following pathways. Reasons for psychological defects caused by restriction of movement.*   15

2. The Horse's Daily Routine
   Feeding: *Standard amount of food consumption. Choice of forage. Harmony of feeding. Utilization of fodder. Period of feeding. Quality of forage. Jealousy. Begging. Stable vices.*   26
   Drinking: *Total water requirement. Quantity and frequency of drinking. Quality of drinking water. Technique of drinking.*   44
   Passing dung and urine: *Dunging ground. Position when passing dung. Position when staling.*   50
   Rest behaviour: *Dozing and dozing periods. Waking stance. Slumber and deep sleep. Rest periods and resting places.*   52
   Solitary grooming: *Function of the skin. The dust, mud and water bath. Rolling, scratching and cleaning.*   61

3. Social Behaviour
   Social grooming   71
   Friendships and enmities   72
   Hierarchy   74

The equine family: *The family structure. The individual family and family clan. Bachelor clubs.*    79
    *Territorial behaviour and greeting ceremony.*    87
    *Special forms of social behaviour. Peculiarities of oriental-type horses. Peculiarities of ram-faced horses. Herd stallion and lead mare. Founding a new family and casting off the herd stallion.*    92

## 4. Sexual Behaviour
Natural mating: *Cycle of heat and natural darkness break. Preheat period. Ovulation. Fertilization.*    101
Covering by hand: *The teaser. Mating.*    110

## 5. Mother-Child Behaviour
Period of pregnancy: *Breed and type variations. External influences. The changed demeanour of in-foal mares. Sign of approaching birth. Birth period.*    112
The birth: *Waxing. Labour. First contact between mother and child. After-birth.*    117
The foal and recognition of identity: *The urge for milk. The suckling expression. Maternal instinct. Bowel cleansing. Characteristics. Recognition of the dam. Frequency of suckling. Weaning. Independent behaviour of the foal.*    120

## 6. Growing Up and Play Behaviour
Incorporation in the social structure    133
Running, catching and fighting games    134
Games with other animals and humans    137
Education from wild to domestic horse    139

## 7. Combat Behaviour and Flight
Fighting games    142
The stallion's method of fighting    143
The mare's method of fighting    146
The critical distance    147
The quadruped test    148
The flight distance    151
Flight and panic    152

# THE HORSE'S FORMS OF EXPRESSION

8. Vocal Expression
   Causes of vocal expression ..... 159
   Vocal contact between dam and foal ..... 160
   The variety of vocal sounds ..... 161
   Audible communication with other animals ..... 163
   The facial expression when neighing ..... 163

9. Facial Expression and Physical Bearing
   Particular elements of expression ..... 165
      *Play of the ears* ..... 165
      *Twitching of nostrils and muzzle* ..... 168
      *Meaningful eye signals* ..... 168
      *Changes of bearing and silhouette* ..... 169
   Orientation ..... 169
      *Assessment by sight and sound* ..... 170
      *Assessment by smell* ..... 172
      *Assessment by taste and touch* ..... 173
   Facial expression of fatigue ..... 173
   Facial expression of greeting ..... 174
      *Greeting ceremonial* ..... 174
      *Neighing to impress* ..... 175
   Mannerisms to impress ..... 176
   Attitudes of aggression and submission ..... 178
      *Threatening attitude* ..... 178
      *Submissive attitude* ..... 180
   Expressions of aversion and lethargy ..... 182
      *The surly face* ..... 182
      *Head-shaking against flies* ..... 182
      *The exhausted face* ..... 183
      *The expression of suffering* ..... 184
      *The look of a mare in labour* ..... 184
   Expressions of pleasure ..... 185
      *Mutual grooming face* ..... 185
      *Playful or cheeky face* ..... 185
      *Mating face* ..... 186

# INTRODUCTION

We humans are very self-centred beings, certain in the knowledge that we are a great deal more intelligent than all other animals. On the grounds of our superior insight it might, therefore, be expected that we would show consideration and understanding towards the animal world, and especially towards those creatures we profess to love and with which we are closely connected. The pains that we take to understand and treat our domestic animals properly, which after all we have made use of for thousands of years and to which we have a lot to be grateful for, can be recognized here and there, but we simply lack the necessary special knowledge. We sometimes fail to show the most elementary understanding, because we do not share the same language. Paradoxically, as natural egocentrics, we demand that our animals learn to understand the human language, and that they understand and react obediently to confusing and difficult demands, which even our own contemporaries sometimes fail to comprehend.

There is no doubt that, after the dog, we love the horse best for a great variety of reasons, which we need not examine here, since they are not always to our credit. It is not inappropriate for a horse-loving author to show a mark of appreciation to all other domestic animals, for they earn our respect in the same way as our favourites, the horses, of which it is frequently said (often completely unobjectively and unthinkingly) how sensitive and intelligent they are. Perhaps it is too much to expect that townspeople, who come into contact only with horses, dogs and canaries, should credit common animals like cattle or pigs with intelligence and feelings. As a serious student of animal behaviour I consider it absolutely unfair to dismiss them completely and to place the spiritual existence of our domestic animals on a much lower plane than that of horses.

Any practical veterinary surgeon who regards an animal as a complex being with a body and soul is daily confronted with the gentleness of a mother sow or the sensitivity of a bull. It is in no way meant as a joke to describe pigs and cattle as sensitive!

It is, however, funny, not to say in bad taste, the way that owners of especially well-bred horses show such appalling ignorance and claim that such 'exquisite' horses are intelligent, sensitive and endowed with almost human feelings. Horses of every breed and type can be sensitive, clever within their scope as animals, loyal, good, kind and whatever else the horse-lover thinks appropriate to his four-footed friend. To label heavy horses and common ponies as basically stupid is just as ignorant as if one refused the 'lower classes' the possibility of possessing a higher intelligence and described them as spiritually dim-witted. No normal person would subscribe to this idea. For this reason, we should not subscribe to such ideas if we want to learn to understand our horses better, and that is the object of studying the behaviour of horses.

Behavioural studies form a fairly new branch of science which for a long time did not receive due attention, but which is now very much the fashion. That serious scientists have regarded this subject with a certain mistrust is due to the fact that within the modern methods of mechanical biological research only small, generally unspectacular discoveries can be made and the fact that the conclusions reached by even the most important behavioural researchers are frequently inaccurate.

Behavioural researchers belong, broadly speaking, to one of two large groups.

The first group comprises the so-called behaviourists, whose research is carried out under an agreed 'scientific' method and principally arranges experiments in order that the animals' intelligence, capability of learning and reactions can be tested. The objective of this series of experiments is frequently stated and every effort is made to achieve this – this method of experimentation naturally produces many individual results, which are valuable if the experimental work has been properly planned. In my opinion this is not always the case, because, in the proper sense of the meaning, horses are not true to type, in that some animals are expected to react to stimuli for which by nature they are not programmed. The results, therefore, of the experiment are considerably more negative than the animals' true capabilities would allow if they had received tests true to type. The positive aspect of this scientific approach to horses is the dummy experiment, in which tests are made to see which optical silhouettes horses recognize as friends, enemies, etc.

The natural method – in which no training is involved – is the course pursued by the second large research group, which primarily concerns itself with so-called field observations to study animals in their natural surroundings without any outside interference. Even this method is not entirely free of mistakes, since the researcher can too easily submerge himself in the subject-matter of these investigations, which leads to the published results not being objective enough. In spite of this, I prefer the second method which, in my opinion, is the only one to appreciate the animal as a whole and to understand it. Apart from the very real danger of the subjectivity of the researcher or student, the danger of humanizing many reactions must also be considered seriously. On the other hand, the behaviour of a large animal cannot be entirely understood without some humanizing, and I believe that Konrad Lorenz, who was the great student and mentor of animal behavioural research, would not have succeeded if he had not carried out his work from the beginning as a real 'friend' of animals in the best meaning of the word. The same objectivity which is necessary in every branch of science is obviously required in all sensible research on behaviour, in spite of the humanizing of working theories.

After so much theorizing let us turn to practical things. Everyone who has long been concerned with horses is in some way either an amateur student of behaviour or a practical person who makes use of such knowledge as he has acquired. If this were not so, we would never physically manage horses, since even the smallest Shetland pony is physically a great deal stronger than most people. The better we can manage our horses, the better it is for both partners. As already mentioned at the beginning of this introduction, understanding is necessary to make the most of the relationship, since, almost without exception, the various horses with which we are concerned are used only as leisure-time companions. It has been frequently said that show jumpers will be successful or racehorses will win only if we force them to do so. Of course this is correct, but it does not preclude a true understanding of the horse's personality. On the contrary, with all the necessary simplification of individuality which is the automatic result of biased training, higher qualifications are permanently achieved only by the correct psychological handling of the horse. Even though such highly qualified horses are sometimes used and shown by riders and whips who have never heard the words psychology or behavioural studies, none the less they make use of them in the correct sense. These successful race and show riders, trotting drivers or cowboys, gauchos or whatever they call themselves, who still work with horses, are natural horsemen who instinctively handle their horses correctly and for whom these horses will respond unconditionally and tirelessly. Most of us have lost this touch through our way of living. We mentioned subjectivity, to which every student of animals is prone, unless he has chosen a simple series of experiments for his work programme; it is this same subjectivity which readers will find equally false or only partly correct when they encounter some of the observations described in this book. Apart from the possibility that even the most careful student can make a mistake, an observation need not necessarily be absolutely true or absolutely false.

Besides the many external temporary factors which influence the behaviour of all living things including people – for example, the condition of the animals under observation, an unusual external stimulus to the optical organs or to those of smell or hearing of the animal concerned, or that the horse is confronted with the frightening proximity of a starry-eyed photographer, or strange-smelling behavioural experts – there are many permanent conditions to consider, which can influence and alter any experimental conclusion that has been reached.

As with people, the horse's behaviour is regulated by the two great components: environment and heredity. Our horses, too, are the product of their surroundings, together with all the circumstances of faulty breeding and training, and the numerous unfortunate experiences, called in modern terms 'frustration', which they have to contend with throughout their lives. Because, in the human sense, horses are more stupid than we are, it does not mean that they are less frequently and intensely frustrated. However their unfortunate experiences and conflicts lie simply, as we shall see, on a plane peculiar to themselves.

Thus the entire method of domestication, no matter how it is done, is a gradual reduction of our horses' freedom and this sometimes produces the same distress psychosis as the frustration of a person who is forced to spend the duration of his life in the culs-de-sac and alleyways of a large city.

I think that one of the cardinal mistakes in our attitude towards horses lies in the fact that, in general, we don't appreciate for example that animals are not created to carry a rider or to run a trotting race, even though they may be specially bred for this purpose. A horse only reacts thus because he is obliged to by people who think it is sensible and useful. As far as the animal is concerned even carrying a rider or picking up a foot, and all the rest of the things we demand, are as completely senseless and incomprehensible as is its food being rationed so that it does not become too fat.

The second component which determines character and behaviour is the hereditary factor. This was denied for far too long, against better judgement. Horses, said official opinion, are by nature good and are spoilt only through bad handling. If horses were secretly removed from stud duties, because they

*Illustrations 1-11*

Plate 1   *above:* Encouraged by his mother's 'funny' rolling position, a young foal wants to play. He shows off and his tail indicates his intention to gallop.
*below:* Love is childish. The in-season Icelandic mare looks with a foal's interest, at the rolling of the stallion, Hrappur.

Plate 2   *top left:* Icelandic gelding Mosi sorts unwanted grasses.
*top right:* Trotter mare Karla Manton shakes off flies.
*bottom left:* Icelandic yearling Skugi paws at grass under the snow.
*bottom right:* Grazing Lipizzaner stallion Conversano Graina appears uninterested but the play of his ears gives the game away.

(Continued page 7)

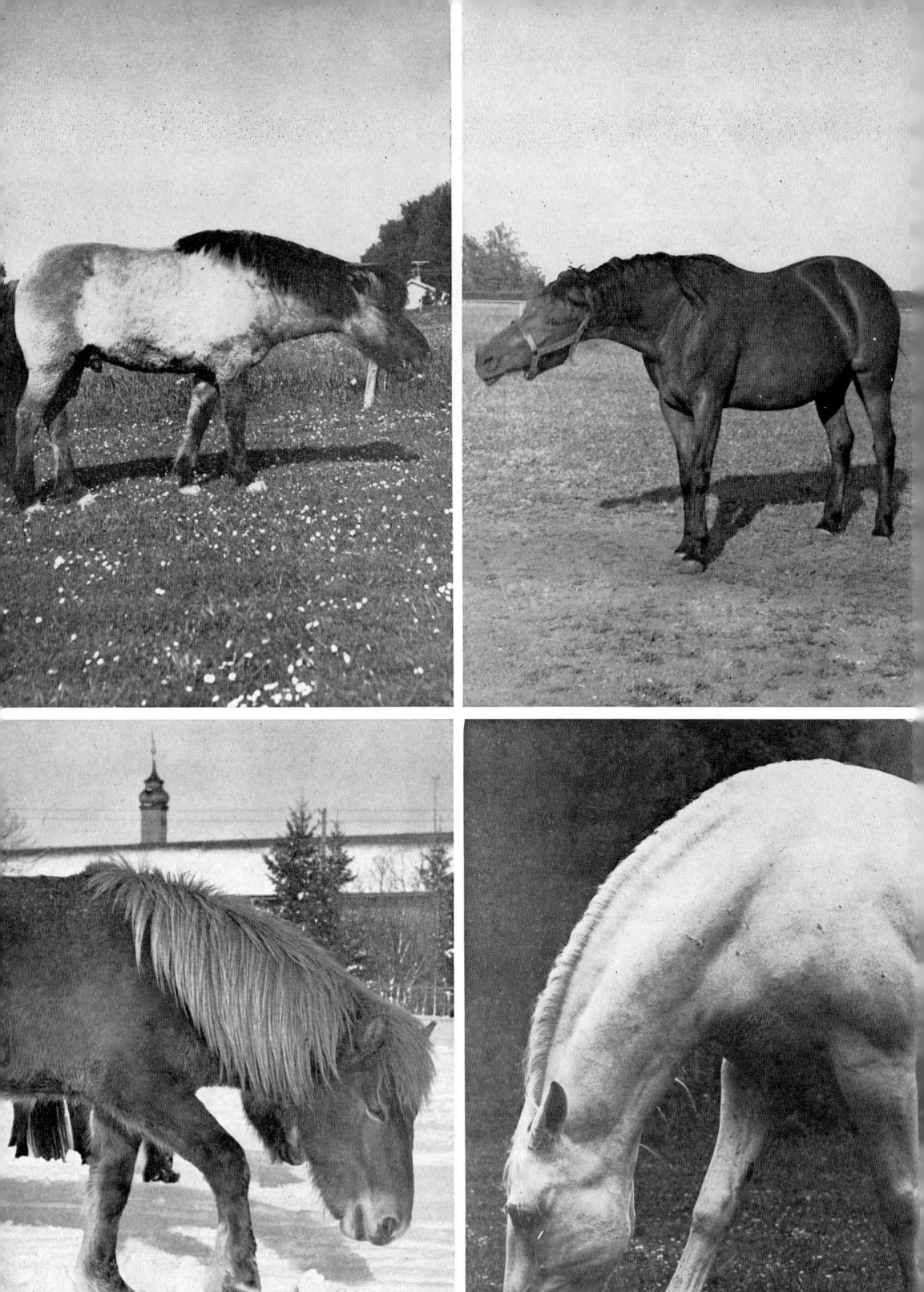

had a nasty temperament in spite of proper treatment, the fact was generally covered up. One is reminded of the reforming advice of some contemporary sociologists who believe that in every wrong-doer there is someone whom the world at large has wronged and, therefore, misbehaviour is an illness. No matter what degree of importance the environment has – and no one would deny that it is important – from earliest childhood, perhaps even at the embryonic stage, the genetic make-up that every individual inherits is, in my opinion, of equal importance.

Since every horse reveals his individual characteristics in his external conformation as well as his temperament, experiments on a few individuals could produce contradictory results. Therefore, the student of behaviour is obliged to choose as large and as varied a number of animals as possible, which will then enable him to find a common denominator. The less the horses concerned have been spoilt by their environment and the longer they have lived under natural conditions, the more accurate the observations will be. Some researchers have, for these reasons, undertaken long journeys to examine the behaviour of so-called primitive horses, which still live in a semi-wild state under more or less normal conditions. The results thus acquired are very important and they have been supplemented by obser-

Plate 3  *above left:* Icelandic gelding Valur staling; his position is typical.
*above right:* Fjord stallion Endo marks a certain spot with a few drops of urine. The lifted tail is an effective signal.
*below:* Young Fjord colts wandering along their track. As the highest-ranking individual, the owner takes the lead.

Plate 4  *above:* Free-range Fjord ponies resting at night, with almost no 'individual distance' between them whilst slumbering; the dozing guard on duty watches the photographer.
*below:* Part of the two Fjord pony families at their communal midday dozing place.

vations made on wild zebras in Africa. Research done on Equidae in zoological gardens showed similar results to those of research on our domestic horses; some scientists rightly refused to have anything to do with these results, because the tests were done on animals living in varying degrees of unnatural conditions. I still consider that these observations and results were valuable, since they helped to provide further basic knowledge of wild animals, simply because tame animals were unable to get away from direct contact with people. In the same way, a thorough knowledge of fundamentals can be used to make correct observations concerning domestic horses.

Inherited behaviour patterns are probably considerably more conservative and slower to evolve than external characteristics. The twelve-inch dog Fifi reacts in much the same way as his jungle forebears – which often makes him look rather amusing – and even we humans have changed very little in our reactions since the Stone Age, except for a few refinements.

If one reads that very interesting book by the well-known behavioural expert Eibl-Eibesfeldt on the Ko bushman society, one is astonished how similar these people on the lowest rung of civilization are to ourselves.

A further factor which, in my opinion, makes research into the behaviour of horses more difficult and sometimes shows very muddled results as soon as one goes into detail lies in the origin of our horses. As I have already proved in other published contributions, our horses are descended from different wild forms, local breeds or sub-species which, in the varying climatic conditions of their once vast Euro-asiatic habitat, adapted themselves not only to different external forms but also to slightly different patterns of behaviour. Our domestic horses carry the blood of many wild forms in varying mixtures and, through selective breeding, certain characteristics are stressed. Thus their behaviour is naturally made up of different components, so that further research does not produce quite such a uniform picture as one would get if one examined, for example, an isolated mountain zebra population.

In this book, therefore, I shall consider those particular differences only when I find it necessary to explain an otherwise incomprehensible behaviour pattern, and I shall confine myself essentially to those characteristics and patterns of behaviour which all domestic horses have in common.

If possible, I shall try to refrain from using too many specialized terms, since this book has no pretention to be a work of science. The object is to help the reader to understand his horse better, to appreciate his means of expression and to understand his wishes.

# THE NATURAL HABITS
OF THE HORSE

# PRELIMINARY REMARKS

In the western world the majority of horses have to live in unnatural circumstances; this includes all working horses which are kept in stables and have to work daily. An exception, up to a point, are brood mares which are out at grass throughout the summer and, to a greater degree, countless feral horses and ponies who have a free range throughout the year. However, even the habits and behaviour of these horses are influenced by humans. In the next chapters we shall try to understand what sort of occupations horses enjoy under natural conditions, so that we can get to know their actual inner rhythm. Then we shall be able to establish how much we disturb this rhythm by using our horses under the saddle or in harness – even stud animals – and how far the various harmful consequences can be reduced or avoided.

# I

# THE PREOCCUPATIONS OF EQUIDAE

All Equidae possess an innate range, rhythm and pattern of movement. What is meant by that? We all know what we mean when we refer to an 'inner clock' by which both humans and animals adjust themselves. At certain times we become hungry and then we eat; at other times we feel tired, so we lie down to sleep. The more regular are our habits, the more punctual is our 'inner clock' – the more, too, we come to rely on it. Horses are known to be animals of habit and routine, to a far greater extent than people, and they keep more precisely to their inner rhythm.

A few years ago it was discovered in addition that Equidae have not only an 'inner clock' – a time/occupation system – but, besides that, their actual activities are bound to fixed places within a definite radius. Accordingly, they not only always feed at about the same time, but they always feed at roughly the same place and they sleep in the same places.

If they are not kept and fed in a loose box, horses live in a given area which they use as we do our homes, with dining room, lounge and bedrooms, or, in other words, they have a range, rhythm and pattern of movement.

That horses do possess this tripartite system to which they adjust themselves within a given sphere of activity is contradictory to a large number of opinions which have been held to date about the carefree life of such highly-specialized, fast-moving animals. In our imaginations a herd of wild horses will cover an immeasurable distance and is not bound by a prescribed preserve.

We will now examine the question of habitat loyalty or the urge to roam, since it is of importance in developing a proper method of keeping horses.

The problem of the size of 'natural' habitat is not so easily solved, since there are few 'genuine' horses living in complete freedom and we have no chance to keep them under observation. In the zoological sense, genuine horses are domesticated horses as well as their wild ancestors, whereas the definition of the word 'Equidae' also includes donkeys, onagers and zebras. The Asiatic or Przevalski horse is, of course, a genuine horse which may have become extinct in his native habitat or at least greatly reduced in numbers, withdrawn and in constant flight from people, so that scientific research into its area of activity is not possible. Apart from this, its rôle as the ancestor of our domestic horses is frequently questioned by many scientists. Yet, of course, crosses with the Asiatic wild horse are fertile, which is not always the case with other Equidae. The wild feral Mustangs and Cimarrons of North and South America, as regards their breeding, are similar to our domestic horses, but even they have become so decimated and, due to constant hunting, so shy that any research into the range of their activities would produce a biased picture.

The next best thing to do is to pay more attention to the so-called primitive horses, those numerous and only partly tamed ponies which often inhabit fairly large reserves. However, even the largest reserve is still too small for research into the area of activity, since long-distance roaming by such animals is eventually impossible because of human population. Probably the study of Icelandic ponies on their native moors would produce interesting results, but, as far as I know, they have only been studied in a German reserve.

The only alternative left is to turn to the relations of our horses who are not so wild and to try to draw conclusions as to their habitat loyalty, change of location or long-distance roaming. The picture is rather confused, as there is only very sparse information on some of the wild Equidae, though about others quite a lot is known. The Harmann zebras, which are fairly loyal to their habitat, are the opposite of some of the Asiatic onagers, which are known to roam great distances. If we establish the cause of habitat loyalty then, for example, the desire to roam will give us a fairly accurate idea of the area of activity of our horses.

Basically it is the ecological requirements – grazing and especially opportunities for watering – which are the decisive factors for every species of animal. The closer and better the quality of the plant growth is, and the closer together the watering places are, the shorter the distance a wild horse or zebra has to travel in order to still its hunger and quench its thirst. In dry areas the number of watering places is the decisive factor governing the area of activity, because all Equidae, however good they are at living in desert areas, still have to drink at regular intervals and have to stay within 40-50 miles of a watering place if they are to have enough time in which to graze.

The unfavourable localities, like those in the steppe deserts of Asia and Africa inhabited by onager and asses, show especially clearly the tremendous reciprocal effect of environment, behaviour and conformation. The great distances between watering places and scarce grazing entail long marches to drink and a sufficiently good set of teeth to cope with the necessary quantity of food in the short time available. Because no enclosed pasture exists, the grazing formation of those Equidae living in this locality is less tight-knit, the distance between individuals is greater and mutual tolerance slighter. The large distances traversed daily entail a body free of fat, and thus extremely fast animals develop a corresponding need for space to gallop around, even in captivity.

On the other hand, all Equidae from well-stocked localities, with rich plant life and sufficient watering places, show less specialization as galloping animals and, since food is near at hand, the distance is smaller between animals and, therefore, they have a distinctly closer social life. The Grant zebras (a sub species of the Burchell zebras) of the Serengeti National Park belong to this group – they were closely observed by Klingel for many years. Although these zebras are loyal to their own locality, they cover considerable distances daily, although they are not especially wide-roaming animals, and their sturdy, round bodies remind one more of ponies than of racehorses. Yet they may cover 16 miles a day, as they do in sunny weather every morning from their sleeping place to their regular, distant grazing grounds (longest measured distance: 8 miles) and return home in the late afternoon. The hour-long grazing, which covers short distances on their regular grounds, is not even included.

Interestingly enough these Grant zebras stay 'at home' during bad weather, which is similar to the behaviour of the feral ponies in the New Forest, England, who remain grazing very near their sleeping quarters when the weather is bitterly cold and wet. There is no contradiction if domestic horses are often very active and gallop around their paddocks when the weather is bad, because it is almost always a case of animals which are not out all day long and want to keep warm.

The weather has a noticeable effect on horses' behaviour, so that it is not surprising if many are weather-conscious. Besides the actual temperature, both the flow of air and the humidity of the atmosphere have an effect on the activity of Equidae, since sultry weather makes horses more lethargic and less capable of working than fine dry weather. Almost all animals react fairly strongly to atmospheric pressures; this is marked by increased nervousness and uneasy movements in thundery conditions, and they do not relax until the storm has passed. Their susceptibility to the *föhn* (the sultry southern wind) which descends from the Alps across the lower ridges with the decrease in atmospheric pressure is even more remarkable. The horses become as irritable as the people who live there. I imagine that horses brought to race at Munich from training stables further afield have won simply because the prevailing *föhn* did not upset their placidity, since it takes some time for one to become susceptible to these sultry winds.

Just as the momentary weather conditions affect the daily activity, so the general climate has a bearing on the radius of movement of horses. Because of climatic conditions and the subsequent growth of vegetation, most Equidae follow fairly well-trodden seasonal paths. In the wide open spaces of the North the winter forces animals to go southwards or at least to seek warm sheltered valleys; in southern zones they are forced by lengthy droughts on long migrations to seek wetter areas, where water holes are still to be found. Even in really ideal localities like the Serengeti, during a drought zebras have to leave the shelter of the huge volcanic crater to cross the hills, which increases their range from 30 to 80 square miles. By comparison, the 3-4 square mile range of New Forest ponies looks rather insignificant.

Apart from scientists whose knowledge is more theoretical, there are practical men who, from experience of the uneasy behaviour of some horses and ponies in autumn, have concluded that such uneasiness is due to a primeval urge to roam. F. Tesio, the great Italian thoroughbred breeder known as 'the wizard of Lake Maggiore', who probably had more classic racehorses than anyone else in the world, made allowances for the roaming urge in his stud programme and in autumn he sent his young thoroughbred horses from Lake Maggiore in north Italy down to the much more southerly pastures in the Roman countryside.

Remarkable practical consequences result from the fact that in wild Equidae not much is normally noticeable of the fleeing steppe runner, since, unless they are actually fleeing to avoid danger, they roam at walking pace.

The walk is naturally the most important gait, which is the reason why exercises at the walk, no matter what the horse is used for, are a great deal better for the animal than chasing it around at the canter or constant trotting.

Since the range of most of the free-living Equidae has not been so exactly studied as that of Grant zebras, perhaps one may draw from the known facts the conclusion that all wild Equidae, and with them also the ancestors of our domestic horses, are fairly faithful to their own habitat as long as the ecological situation does not force them to roam great distances. This habitat loyalty, which in any case always includes an area of several square miles and in no way resembles an ordinary animal reserve, is not only characteristic of short-range zebra species but also of the typically running animals like onager and kulan, whose change of habitat is adjusted to the available watering places.

An ancient phenomenon should be mentioned, before we turn to the 'symptoms of mental deficiency' in our domestic horses which are often the result of their miserable radius of activity: all Equidae cultivate walking in 'crocodile' from their sleeping quarters to pasture or from thence to the watering places. The remarkable thing about this follow-my-leader is – and for the moment we will leave out the order of precedence of the members of the herd – that the same path is always used, no matter whether it is in a narrow, hilly region with a number of suitable walking places or whether it is open steppe where any number of animals could walk abreast. The resultant small pathways, which criss-cross the range of a group of Equidae in all directions and which are easily seen from the air in an aeroplane or helicopter, are found on every paddock, no matter how small, where domestic horses graze. The inclination to make pathways or roads appears to be one of the primitive urges of higher mammals, because it is found in most wild and domestic animals, even humans. One need only observe how children love to trample a path in a meadow. Even adults keep to a known road just as uniformly and stubbornly as horses do in order to reach a required destination, without stopping to consider if there might be an easier way. In the case of wild Equidae which today inhabit open grass, tree or bush country this custom is a legacy from millions of years ago, when their ancestors were only fox-sized, many-toed creatures of the undergrowth, which lived in the thick jungle of tropical forests and which, like our wild animals, had to make pathways in order to progress faster.

If we consider how much exercise a wild horse takes in the course of a day, how many of his own kind and other animals he will meet, and what different places he will visit, with the many changing impressions of vision, hearing and smell, the happy effect on our horses of being ridden or driven in attractive country cannot be overestimated. Since even Equidae with a very localized habitat loyalty enjoy a variety of new impressions each day, it is wrong to think that only thoroughbred horses need plenty of exercise and a change of scenery. In fact all horses, whether heavy horses, warm-blooded horses or polo ponies, need a constant change of interest to keep them temperamentally well balanced. Horses react differently according to type and, although a reaction may not be observed, we know that they require stimulation.

A well-known reaction, which arises from the dull life led by many highly capable horses, is the so-called sourness. Some psychologically weak horses are unable to bear the lack of variety in their working day or the dull routine of a training programme. They become sour – that is to say, they react unwillingly to work, cannot digest their fodder and generally lose condition. Dressage horses lose their elasticity and their equilibrium, as we saw with the Russian Olympic horses, which had, in my opinion, lost all their pleasure in movement and some of which appeared to me to be completely apathetic. Show jumpers often refuse fences which they once took in their stride, or they jump carelessly. Some racehorses break into a sweat; Trotters jump about or trot in false tact. Thus one of Germany's most successful Trotters, Simmerl, could only be trained at home on his own farm, because he simply could not cope with the noise and bustle of the race course. He was homesick, ate badly and showed so little ability that one would never have believed what a wonderful horse he really was. It was only after a holiday at home on his own paddock that he lost his psychoses and his physical ability returned.

In racing, the primary concern is money, while in show jumping it is the satisfaction of personal ambition or sometimes perhaps a compensation for one's own complexes. Sports demanding great endeavour are, in every sense of the meaning, carried out on the back of the horse and it would be superfluous to agree with the cry of the sentimental non-riders or fearful

*Illustrations 12-21*

Plate 5 *above*: Fjord pony foals in a deep sleep.
        *below*: New-born warm-blooded foal Peppoli sleeping like a dog.

Plate 6   Icelandic mare Gledia about to lie down, then rolling and getting up; her day-old foal is puzzled, then interested and finally animated into little jumps.

(Continued page 23)

horsemen that this is a form of cruelty to animals. At a time when extra attention is being paid to the quality of life, it is obvious that any cruelty should be kept to a minimum. Generally speaking, neither the relatively few event riders nor the weekend horsemen are wicked sadists; it is the so-called horse-lovers, sometimes out of a false love of animals or lack of knowledge of proper equine management and possibly out of pure laziness, who are just as cruel to animals as the much publicized event riders. Their form of cruelty, which is not actively committed and is of a more passive nature, does less damage to the physical character of the animal than to its psychological well-being.

We have already mentioned the form of imprisonment, twenty-three hours of solitary confinement, which is the lot of some of our pleasure horses. If one remembers the horse's basic love of freedom, this is a case of cruelty to animals, however hard this may sound, and is not to be disguised by painfully clean stable yards, luxury boxes and even the excellent condition of the animals themselves. Paradoxically, as far as

Plate 7 *above:* Intensive coat-nibbling forges strong social links between horses. The stallion Findo nibbles one of his mares (*left*). Proof of the astonishing fact that two herd stallions can live together in absolute peace. Mutual skin care between Endo and Findo (*right*).
*below:* Single family – right the mare, in the centre the stallion Findo and, near him, the foal and the thin yearling, which in spite of his poor condition is not continuously eating.

Plate 8 Battle for hierarchy
*above:* The stranger, the black Trotter mare Onda, threatens the warm-blooded mares Ganda (*centre*) and Pepita (*right*). The hitherto lead mare, Ganda, tries to impress with her bent neck, tail carriage and passage action, but her ears already show her submission. Pepita tries, with tail and hindquarters retracted, and submissive position of the ears, to get out of the way.
*below:* Onda positions herself to kick; notice her threatening expression. The warm-blooded mares take flight.

daily excercise is concerned, it is the spoilt, privately-owned horses which suffer most. The 'hirelings', which are called 'school horses' in fashionable riding schools, have to go out several hours daily, although generally they will be properly exercised for only an hour. In bad weather they will be exercised in the school, so that some horses for most of the year rarely see the outside world. From lack of change and new impressions such animals must become mentally and physically stunted – there are plenty of parallels in human psychology. The results are stable vices like walking the box or weaving, which very clearly reveal a lack of exercise. Lively horses with a marked desire for exercise understandably easily fall victim.

In the case of walking the box or weaving one can describe it as stereotyped exercise. Some horses ceaselessly march round and turn constant circles or, if there is room, figures of eight. These stereotyped movements, which we have all seen in the great cats confined to their cages or Equidae in zoo enclosures, are not purposeless but are carefully judged, as one can recognize by the fact that the head is held in a definite direction. Weaving is generally done by animals which are kept tied up in stalls and have not even got the limited freedom of a box. Since they cannot turn round, they have to satisfy their urge for exercise by a constant sideways movement, lifting alternate forefeet, and changing weight from one foot to the other. The repeated hour-long swinging from left to right of the head, neck and forehead reminds one of the way a weaver's shuttle shoots backwards and forwards, and in its monotony there is something depressing and confused. If these horses are put in a box, they continue to weave, as they also do if put in a yard without grazing. They do themselves little damage and pulled tendons are rare, but the constant movement requires energy and, like many vices, weaving encourages other horses to imitate, especially those of nervous temperaments.

From the horse's point of view the weekly rest day is nonsense and typical of the false notions emanating from human attitudes and love of animals. With praiseworthy intention we believe that our pets, like ourselves, need to relax, which, when they were working six days a week, was necessary. Even then the tired animals had some sort of pasture in front of the stables and most farmers gave their horses the chance to have their harness off and to roll and clean themselves, and to while away a few hours as they felt inclined. A. Paalmann, who was the successful national trainer of the British jumping team, would not allow any horses in full training to have a rest day – to avoid boredom, he developed an elaborate system of training in which there was an alternation of dressage, jumping in the school and in the open, ridden exercises, lungeing, free jumping and periods of grazing – rest days were consequently not necessary.

Unfortunately most of us do not have the opportunity to give our horses alternating exercises, but we can try to make the best of what we have. In the country, these problems are fairly easily solved with a paddock or yard, if one cannot find time to exercise the animal. In the town, perhaps the horse can sometimes have free run of the school or a longer exercise period under the saddle, because, in contrast to getting a horse fit for competitive work, when the most important thing is the quality of the exercise and systematic training, any animal's psychological well-being depends on the duration of his opportunities to exercise himself. If we want to be kind to our horse, it is better that we let him amble along with us for several hours, with a short trot or canter now and then, than to start by galloping here and there, and then to put him through his paces, because we happen to have time for it and because we think that he has the rest of the week in which to be idle.

# 2

# THE HORSE'S DAILY ROUTINE

In order to understand and tolerate the behavioural habits of our four-footed friends, it may be interesting to follow the routine of a horse's day under normal conditions. We shall not achieve any sensational results, but perhaps some of the characteristics which we suppose to be bad will prove to be the exact opposite and will appear, in the wild, as a sensible action and reaction.

## FEEDING

The most important factor in the life of a horse is feeding. Possibly this sounds rather shocking to the ears of admirers of the noble horse, which appears to personify elegance, courage and fire. However, eating or feeding is the most important activity in the life of a normal horse as it is for every other living creature, because, without sufficient foodstuffs, there could be no individual life and no continuation of the species. This essential occupation, which naturally is of equal benefit to man and animal, has become one of the most time-consuming daily activities of Equidae due to certain anatomical peculiarities. Horses are, as we know, herbivorous. Since even high-grade plant fodder is never as concentrated as animal protein, they need a relatively large and bulky quantity of forage in order to retain their body weight under normal activity (maintenance fodder), i.e. to guarantee growth of the embryo and foals (in freedom, a form of production fodder). The horse's entire digestive system is, therefore, quite comprehensive and able to make the best possible use of poor-quality forage. In order to give the digestive juices and the intestinal organisms the necessary 'fighting material', coarse and rough fodder must be exceptionally well ground. This problem, which is common to all herbivores, is solved in different ways. The large group of ruminants including cattle, sheep, antelopes, camels and deer, which swallow quantities of fodder fairly rapidly, help their digestion by chewing the cud, generally when resting; they bring up the chyme into the oral cavity and chew it carefully. For Equidae, which do not possess a third stomach but have instead a tremendously-developed appendix that deals with the main processing of food, it is important right from the start to have a good mechanism to break up the food in one simple operation.

In spite of their larger and broader molar teeth, which work like millstones, horses eat much more slowly and carefully than ruminants, but they pick out single grasses and plants with more accuracy than cattle do, perhaps because of their careful and relaxed method of chewing. The selection of forage is interrupted by definite pauses, although, if they are very hungry, they are not so choosy; even so, really hungry horses do not just swallow anything, unlike cattle, which will swallow inedible materials such as nails and wire, sometimes with fatal results. Horses pick grasses with their lips and then bite them through with their incisors, with a backwards-upwards jerk of the head. The denser and more variegated is the plant growth, the more carefully the single grasses are chosen, and it is astonishing how easily a horse with his fairly broad muzzle bypasses less tasty herbs and how cleverly he drops ill-chosen grasses from his mouth. Nails and other rubbish under the oats in the manger are carefully sorted out and left after feeding, even by hasty feeders.

Sometimes, however, horses have no other choice but to eat unsuitable foreign matter, if the whole feed is not to be rejected. If very dirty, green fodder is put in the box then the earth clinging to it will be eaten, which in some circumstances can lead to diarrhoea. On very bare pastures horses pull out the roots with the grass and, naturally, earth still sticking to the roots cannot be separated. If the soil is sandy and the animal spends long periods grazing, the incisors and molars will soon become worn, as sand is an active grindstone. I noticed in Portugal, in some particularly poor districts, how all the mares which grazed these pastures, which were dried out for most of the year, had to be culled from stud when they were between eighteen and twenty years old, because their teeth were completely worn away by the quantity of quartz sand they had eaten.

For anatomical and ecological reasons this time-consuming method of feeding varies amongst Equidae, and yet on the whole it is very like a programmed event on a daily time-table, which I shall call fodder time. The innate urge to feed at a predetermined time during the day is so deeply-rooted in all Equidae that they try to keep to it even under unnatural physiological conditions. When the enclosures of zoo Equidae show no sign of plant growth, the animals will wander with their lips to the ground and make movements with their lips as if grass were actually growing.

Two remarkably strange examples demonstrate how stubbornly our horses are bound by their fodder time. It can happen that horses in a starved and very poor condition will simply stand for hours in the middle of excellent grazing, although they would benefit if they fed off it. On the other hand, some owners are considerably annoyed by the apparent greed of their well-fed horses. They take great exception when these horses, which are not lacking in either oats or hay, start eating their straw beds and develop big bellies. Both observations – the horses which refuse to graze and those which apparently senselessly eat too much – are easily explained by the above-mentioned programming to a natural, basic fodder time.

From my own observations, adult domestic horses spend an average of twelve hours a day feeding; seasonal influences and a varying amount of available forage may shift the time-table of free-ranging horses slightly, but these factors cannot appreciably lengthen or shorten the duration. With regard to the number of hours, it seemed to me that slight variations occurred with different breeds, although I will not absolutely vouch for the rather longer period, which I surmise may occur among the northern ponies. We over-civilized western Europeans may think that twelve hours daily feeding is rather plebeian, since we are more than ready to transfer our silly human fashions, by which we currently set such store and which insist that only a slim appearance is attractive, to our innocent horses, which most certainly do not find a belly like a gutted herring and an obvious rib cage particularly desirable attributes.

Shocking as the twelve-hour feeding period may appear, it is completely sensible and is not a practice limited simply to the much-despised common horses. The healthy attempt to be fat if possible is Nature's way of guaranteeing the survival of the individual animal and that of the species, when the good months with their plentiful supply of fodder succeed in making the hard months easier. All the free-ranging members of the Equidae species must in time, therefore, put on a good layer of fat on which they can live when times are bad. It is a matter of surviving frost and snow in the cold regions of the temperate zones to which the larger part of the range of distribution of the genuine wild horses belongs; in the hot regions, where zebra, wild asses and some onager live, it is the drought which causes the green grass to dry out and become nothing more than ballast. Those Equidae which do not range far from home to seek fresh pastures, when forage is short on their home range, are remarkable for their eagerness to feed, and at the end of a good season they have acquired a fairly substantial layer of fat, without which, for example, they would not survive the raw Siberian winter or the dry summer of the desert steppes.

The natural urge to gain a layer of protective fat is not limited alone to animals. Even for us humans, if we are not too spoilt, a natural pleasure in good and sufficient food is absolutely proper. One should not be deceived by the modern idea of what constitutes healthy living or by the publicized ideal of beauty – they are quite recent and are by no means second nature. No one can make me believe that Man's common urge to eat is simply a matter of compensation, due to some spiritual deficiency. We should remember that even Europeans have suffered very frequently from good and bad years, which fortunately is no longer the case in North America and western Europe today. In addition to this, we are not programmed for a sedentary life, for Nature created us as extremely active, fleet-footed animals. Our unnatural existence, our self-domestication have, however, not been able to extinguish the inherited, age-old, specifically human 'greed'; it is not without good cause that dieting and maintaining a slim figure for aesthetic or health reasons are often very hard. A normal person likes eating as much and as plentifully as did his forefather in the Stone Age, and for this reason the ideal figure of many primitive peoples and of the underdeveloped countries is a buxom type and not the person in 'racing' condition.

Even horses in racing condition, magnificent and elegant though they are, are not sufficiently nourished in relation to their original standard of fodder consumption. I am sure that some of those weaknesses of temperament which are frequently repeated in racehorses and racing Trotters originated in the frustration of these animals over many years in their urge to eat more than they could get, and that crib-biting, wind-sucking and tongue-swallowing are vices probably resulting from an unsatisfied hunger, which they have had from foalhood. Without doubt, racehorses in training are fed on the most nourishing food and are brought to top physical condition, and in that sense they certainly do not suffer from hunger, which chiefly originates in the body cells, when the concentration of staple nutrients becomes too low. Even hunger which comes simply from an empty belly is not present when such horses are properly managed, since they get regular feeds of highly-concentrated fodder and their digestive systems have grown accustomed to little roughage, and, therefore, the smaller quantities are absolutely sufficient to satisfy them. However, I am doubtful whether this method of feeding is sufficient to satisfy their natural psychological needs, because even well-bred Thoroughbred horses are endowed by Nature with the same urge for a standard amount of food as their less wiry brethren and if they did not have to suffer the kind of forced feeding which is deemed necessary for a racehorse then their conformation would be more natural and more like that of an ordinary warm-blooded horse. The horror exhibited by many of the Thoroughbred horses' best friends when they see their favourites later at stud, in the conditions normal for breeding, is characteristic of the current ignorance of the essential biology of the horse.

How then is it possible that animals of the same age, raised in exactly the same conditions, grazing the same pastures and in winter receiving the same hay and corn ration, are often in different states of condition, which result in the animal being more or less active? In the horse-breeding world this phenomenon is generally explained as the horse being a 'good' or 'bad doer' – a few years ago the prevalent belief was that it was the soil on which the horse was raised which was right or too poor for a certain type of horse – but this was only a symptom and not a cause. In my opinion, the cause lies in the diverse origins of individual breeds of horses, which inhabited a variety of *milieux*, under multifarious conditions with quite special data, and from which characteristics developed which as conservative Equidae they have retained. In this connection it is much the same whether one assumes, as I do, a polyphyletic derivation for our horses or whether one believes that they originate from a single wild species and have developed diverse characteristics under the influence of humans in their respective geographical localities during the 4,000 years or so of continuous domesticity. There is no disputing the fact that Arabs, Turkmenes, Berbers and Andalusians, which represent horses of southern zones, have developed physiological and psychological characteristics or features different to those of northern European or northern Asiatic cobs and ponies. On the assumption that all Equidae need approximately the same amount of food – the supposition being based, among other things, on literary references to genuine wild horses living in freedom – it is of paramount importance to the nutritional condition of a horse at grass that he exists in surroundings adequate for the needs of his particular breed. If his surroundings are adequate, for example if we have given him a grass paddock which is large enough, a healthy horse *will* show good condition in summer, without additional feeding, since his grazing period corresponds to the nutritional value of the grass. If the grazing is too rich for a breed of horse which naturally belongs in surroundings with poor pasture, then in a very short time all the animals become fat, since they still graze the normal number of hours and thus eat too much of the rich grass.

To become too fat on average pasture is a recognized fault

of the so-called cob-type horses: the Haflingers, Icelandic ponies, Fjord, etc. These animals are by no means greedy, as one often hears, but, because they have excellent teeth and good digestion as well, which is more than able to turn even poor-quality forage into optimal nourishment, they eat far too much good grass during their twelve-hour grazing period. These breeds do not belong to our climate, at least not in spring or summer, and sometimes they get so fat that the fertility of the mares is affected. On the other hand, some horses do badly even on fairly good pasture. Given that the animals are healthy, that their teeth are in good order and that they do not require de-worming, the horses which do poorly are not getting enough nourishment from their standard amount of food. In our environment they need additional fodder. The chief factor in this standard intake is simply that an almost exact number of grazing hours is adhered to, whether the fodder is adequate for the animal, too abundant or insufficient.

Horses which have a smaller surface to their teeth, like some well-bred Arabs and Thoroughbreds, even with twelve hours of grazing in certain districts are for purely mechanical reasons unable to chew enough grass and, therefore, to digest it. They do not graze longer hours to make up the deficit, instead they adhere to their rest periods exactly as do their well-fed or obese grazing companions. In addition the vegetation is often too wet and of a quality too poor for these horses. The following example from Spain shows just how true this can be. Southern Andalusia has the same sort of arid climate as North Africa and North Arabia, and is marked by only a short period of vegetation in spring, when there is rich and nutritious grazing, and by a long, and exceptionally poor, dry period. Whilst the indigenous, pure-bred, ram-faced Andalusian mares, which are kept out at grass the whole year round, are noticeably fat in late spring and are still reasonably well covered even in early autumn, the condition of cross-bred horses with a large amount of Thoroughbred or Arab blood, and that of pure Arabian mares, becomes so poor during the drought and bitterly cold winter that they do not really recover during the short spring months. Traditionally only the stallions were given additional fodder and the mares were kept exclusively out at grass, receiving only a little extra wheat straw. It became questionable whether it was economical in the South Iberian peninsula to keep animals with small Arab-type heads and jaws, whose ability to masticate was not so efficient, therefore, and whose intake of fodder was quite out of proportion to the fibrous, un-nutritious food available.

In general, stabled horses are given not only the amount and kind of food prescribed for them by their owners, but are also fed regularly. Normally they are fed very early in the morning, at midday they are given a smaller ration, and the evening feed is given about 3 pm. Let us consider whether this is just practical for the groom or whether this system equates with the normal and natural habits of horses enjoying a free range.

Literary references to the grazing periods of wild Equidae are rather varied. Perhaps this is because different kinds are referred to and they have slightly different daytime rhythms. The locality, too, of the particular equine herd, with all the ecological factors, season, temperature and possible uneasiness due to an enemy, of course, are all component factors in a horse's day. A most important element which affects the conclusions is, to my mind, the person who is doing the observing. Only a small number of people have watched a herd for a full twenty-four hours without interruption and then they have usually confined their investigation to a single session, so that possible disturbance by outside influences could not be taken into consideration. What is clearly apparent is the fact that Equidae enjoy many hours of feeding, sleeping and superficial rest, and it may interest the onlooker to know that they never sleep throughout the night, as one might expect of diurnal animals – but I shall return later to the subject of sleeping habits. First I would like to give an account of my own observations of the hourly feeding times of stabled horses under different methods of stable management.

Two socially complete groups of Fjord ponies, consisting of thirty-four animals, which occupied a twelve-acre enclosure during the summer months, enjoyed two long and two shorter feeding periods in twenty-four hours. The first main feeding time was taken by both herds, apart from one another, not long after dawn, beginning between 4 and 5 am and lasting until about 7.30 am. Between 11 and 12 am there was another short period of grazing, and a second short period of grazing between 3 and 4 pm. The fourth grazing period began in the late afternoon from 5 pm and lasted, broken only by one or two individual animals, until midnight. I watched them for a number of days during the summer. If these periods are added together then we arrive at the twelve-hour feeding period already mentioned, which other authors also discovered with different herds of Equidae. With my own herd of twenty warm-blooded and Trotter mares out at grass in midsummer, the long, morning feeding period was considerably extended, and the mid-morning feed dropped, so that they generally fed through until the flies began to plague them at 10 am.

My observations during the winter months were made mostly during daylight hours, since I must admit it was too cold for me to be outside at night. However, even in winter, the daily routine of the Fjord ponies scarcely changed from that of the summer and was only slightly influenced by the addition of a little hay. The customary times of feeding stabled horses appear in principle to agree with the horse's natural grazing periods since, apart from the type and amount of fodder, the main feeding period is given at night and the smaller midday feed fits into a well-organized daytime routine. Since the owners of big horses and most of the pony-breeders with their open-range stable-management generally have a far greater influence on the habits of their animals, one should not dismiss this fact as unimportant to our traditional stable management.

It is interesting to note that my own horses, which spend most of their time stabled in winter, still maintain their daily twelve-hour feeding periods. Since my method of feeding is in some ways different to the general one, observers would find it difficult to check this. It should be noted that I am writing only of my own brood mares and their offspring – horses which have a breeding and growth rate to achieve, but are neither ridden nor driven throughout the year. Besides abundant pasturage in summer and in winter a daily period of several hours outside, my stable management is unusual for the fact that the pasture, so to speak, is brought into the stable. To the horror of most visitors my horses always have access to green forage of various kinds, so that, as with their summer day and night grazing, they can feed in the stable when and how they like. The average person finds them too well-fed and with too much belly, but I think that, as far as nerves are concerned, they are considerably more balanced and have far fewer stable vices than horses of a more elegant shape. The drawback to my feeding method is that it needs an expert to recognize the quality of such animals. To be able to judge the quality of stud animals in what I call natural breeding conditions requires just as much knowledge as it does to discover the qualities of a starving horse in bad condition.

There is a great deal of discussion about the variety of fodder but this is seldom put into practice. Free-range horses, which can graze over a large area with different soil and plant life, by no means feed on the same grasses and plants all day long.

Thus, from my own research I can concur with Ebhardt's observations on Icelandic ponies that these animals, too, in the early morning and the midday grazing periods, graze from good, rich, sweet grasses on the higher sunny slopes in order to get their main nourishment. During the short afternoon period and in the early evening hours, before they reach their sleeping quarters, they graze only in the lower, damper valleys, where the grasses are rather sour. Some Icelandic and even Camargue horses are said to stand in swamps in order to browse off the underwater sedge shoots and thus to achieve a dietetic reaction which helps the digestion. I am inclined to believe this, since nothing seems to contradict it, which is why the winter stable forage follows this natural sequence of choice of plants in giving the horses the best hay in the morning and, to freshen it up, sour meadow hay in the evening.

Because of their special digestive organs, horses need a certain amount of roughage and cannot exist on concentrates alone. Although it contains all the minerals necessary, a good all-round diet also includes bulk forage. Even racehorses receive a small amount of hay to keep their digestion in order, but sometimes, if they are not getting enough, their need for roughage becomes apparent when they start eating dirty straw. I scarcely need to repeat that, if a horse in training has to wear a muzzle or is put on wood chips or peat, he becomes badly frustrated. In livery stables where concentrates are limited but plenty of straw is used, although the horses do not look in such good condition, they are in fact physically healthier than many corn-fed luxury animals.

Some times horses eat earth or lick and nibble at the stable wall, which means that they are lacking in minerals or salt, and, in spite of additions of both these trace elements, they continue this habit. In any case in modern stable management trace elements, minerals and vitamins and a salt-lick are part of the horse's normal diet. The latter is necessary, as with ruminants, if green forage with its high potassium content is to be properly digested. This book is, however, not the place to continue a discussion on the values of trace elements.

In spite of that, however, I would like to mention that many horses enjoy certain leaves, barks or herbs. Ebhardt points out that free-range Icelandic ponies feed off selected medicinal plants besides the usual grasses, in order to avoid worm infections and other illnesses. Although they obviously would not understand why they were eating them, I believe that not only the hardy strains which live naturally but all horses, in any kind of environment suited to them, have retained an instinct for the necessary, or simply suitable, herbal medicines. This opinion is strengthened by one of my own experiences. On their way to the paddock our horses have to cross a yard in which some large horse-chestnut trees grow. When they return in the evening, two groups make for the chestnut leaves. The first group consists of old and even very old mares, of whom one suffers slightly from heart trouble; the second group consists of brood mares of various ages and stages of pregnancy. It is fairly certain that these horses are much livelier after eating a quantity of the leaves than they are on the days when they have no oppor-

*Illustrations 22-31*

Plate 9   Taboo ceremonial: After initially showing off, Findo and Endo smell each other on the shoulders, on the flanks and on the dock.

Plate 10  Ritual territorial behaviour.
    *top:* Endo dungs ostentatiously, while Findo smells his hindquarters.
    *centre:* After both have turned on the forehand, they smell the heap of dung together.
    *bottom:* Nose contact and eventual parting.
(Continued page 39)

tunity to do so. Horse chestnuts contain a well-known preparation used in human medicine to combat cramp and to dilate the veins. This quite definite feeding on the horse-chestnut leaves, particularly by old and in-foal mares, who, like people, suffer from contraction of the arteries, is another example which bears out Ebhardt's observations.

The healthy feeding instinct, already described, seems to contradict the apparent frequent greed of certain animals, for instance those that manage to get to the oat bin, but in fact it is just the contrary. On principle one must first remember that the oats which we are accustomed to feeding to horses in work or training are actually too rich, particularly in the calorie content of the varieties imported from overseas, and that they are not a natural fodder for horses. Horses of all types are primarily grazing animals and, as already explained, the nutrient content of grasses on the former ranges of wild Equidae was extremely varied. Secondly one must remember that they eat herbs and leaves and, in their original habitat, corn was available only in small quantities and at certain times of the year. All forms of Equidae show the same preference for 'nature's oats' growing wild, the seeds of individual grasses which contain almost the same nutrients as our specially-selected cereals. Zebras nibble the seed heads of long grasses as they go on their way from their sleeping quarters to their daily pasturage and, if we gave our

Plate 11  *above:* The colt 'bachelor club' sniffs interestedly at the newly dropped dung of the young mare.
*below:* Practising territorial behaviour with aggressive showing off by the stamping four-year-old, which is not answered by the dark faced, higher ranking Asko.

Plate 12  *above:* Endo and Findo approach the gelding, which is a newcomer.
*below:* Old and young stallions co-operate against the intruder; the gelding tries to ward off the four stallions.

stabled horses the chance to be turned out on a pasture with high and overgrown grass, like wild zebras, they would pull at the grass heads and panicles as they wandered.

In their natural surroundings, therefore, horses obtained only a limited amount of corn to eat at any one time and perhaps this is the reason why many of them vehemently push their oats out of the manger, since even today they prefer to eat single grains. Grazing horses fill their relatively small stomachs (maximum 32 pints capacity) several times a day, as the easily-digested food for a normal meal is constantly pushed further into the adjoining bowel. So a horse eating out of the corn bin feeds as if he is grazing normally, without stopping, and only ceases when his stomach is full. In a very short time, the chewed oats expand considerably and this results in a painful dilation of the stomach, causing a spasmodic closure of the digestive tract and blocking the normal transmission of the contents into the small intestine. For anatomical reasons horses cannot be sick, so the stomach cannot be relieved by this means and under certain conditions the stomach wall can split.

Some horses have a strange and unnatural preference for particular kinds of fodder. It is well known that the herbivorous horse can become accustomed to eating animal protein mixed into its food, like fish meal or egg and milk powder; even Tibetan post horses will eat fresh sheep's blood mixed with millet gruel, and Icelandic and Shetland ponies, when really hungry, eat seaweed and fish offal. Raw egg is frequently used to get horses in top condition (and stallions during a heavy stud season). I had a unique experience with a young warm-blooded mare which was passionately fond of eating dead day-old chicks that she cunningly stole from our numerous cats and which she then ate – beak, claws, feathers and all – with every sign of enjoyment!

A more objectionable habit is that of young foals which eat

their dam's dung apparently with relish. Apart from the fact that they thereby infect themselves with worm parasites at a very early age, this behaviour appears to be physiological, since it is thought that the foals find in the dung the intestinal organisms which will be necessary for them later on. Probably it is not so well known that a bad case of thread worms, which often occurs in foals and young horses, is responsible for extreme nervous disturbances, as the worms secrete a poison injurious to the nerves. The outward symptoms of exaggerated fear, hectic conduct and odd, incalculable reactions can be quickly reduced if a worm-cure is undertaken in time.

With free-range grazing the choice of forage is, of course, regulated to the seasons. Thus, in the autumn Icelandic ponies are so greedy for acorns that one can almost compare them to wild boar, which fatten on acorns; the acorns provide the ponies with a layer of fat, which protects them from the cold of the approaching winter. Later, the animals turn to twigs and tree bark, and even include conifer branches in their diet. In winter food usually has to be pawed from under the snow; in contrast to their action when cantering, special use of either the off- or near-side foot is not noticeable. Some horses rummage in the snow with their muzzles, looking for grass, and then push the snow aside like a snow-plough. Horses from the northern countries have developed a specially-shaped nose and muzzle. It is interesting to note that zebras, too, in southern Germany look for food under the snow, which shows that pawing the ground is phylogenetically a very old habit. We shall return to this later.

When horses are stabled, pawing can become a troublesome and noisy vice; usually this occurs just before feeding, when many animals become extremely excited and demonstrate their enormous hunger by apparently threatening gestures. Under natural conditions, jealousy, which can take dangerous forms in some horses, is never possible, since Nature supplies a well-decked table for all members of the herd alike; thus there is never any necessity to fight for food as do animals of prey. During the long periods of grazing, horses build up distances between each other, although the family bond remains quite secure, but they can also suddenly find themselves grazing close together without feeling mutually threatened. However, as soon as there is a shortage of fodder, they behave as humans do with a comparable healthy egoism and jealousy, and the strongest animals make sure of their share first. One might almost suppose that such obviously strong jealousy in the ram-faced horses is explained by the extremely poor habitat from which these breeds originally came, where their ancestors had to defend every tuft of grass. The consequence of this behaviour, which is simply due to a restriction of food, is that horses of this type show a marked restraint towards each other.

Jealousy amongst stabled horses is common, because, unlike my own animals, they do not have access to forage throughout the day and the relatively small, regular feeds are insufficient. This is clearly seen when one puts grass or hay into a yard: the horses of highest rank drive all the others away in great style, so that not only is the fodder spoilt but it is also shared very unevenly. A similar situation can be seen in many studs when the foals are weaned to share two to a box. This excellent idea, which prevents the foals from becoming lonely, produces the disadvantage that often one foal does not get enough to eat if they are not tied up at feeding time, since in this situation too the stronger foal will drive the other away and the latter does not do so well.

An awkward habit amongst many horses, which they learn only from people, is to beg for food. Many people cannot pass a meadow without giving their own or even strange horses a lump of sugar or a carrot. Although this might seem to be a friendly gesture, it is not good for the horses. If one approaches the paddock fence armed with a lump of sugar, there is a rush to be first and the higher ranking horse snaps up the tit-bit. Obviously this annoys the other horses, which would also like some sugar, and they then proceed to show their displeasure by biting and kicking their unfortunate lesser companions, which are waiting at the back of the group. Just as obviously, someone will be hurt. Such silly familiarity has the result that no ordinary person can go into the paddock without almost certainly being knocked over in the dispute which greets him. If the visitor is hurt, the subsequently howls of pain generally mean that the animal responsible is labelled as dangerous. My own horses never get sugar and I insist that they come to me because they want to, not in the hope of getting something. Visitors' lumps of sugar find their way into our coffee, as long as they do not come from little boys' trouser pockets! One may want to give one's horses tit-bits, if one can avoid a scramble, and the stable is the best place. Carrots are better than sugar and contrary to popular belief do not carry parasites; best of all though are apples, which all horses love and which they will pick up from the ground or off the branches of trees if they can be reached.

Foals have to get to know their surroundings using all their senses and their chief method of discovery is, like that of little children, by their sense of taste – they 'put everything into their mouths' – they nibble and lick everything that is unknown to them. However, adult horses usually bite at anything handy out of sheer boredom. This occurs because their feeding periods are too short. Active horses short of work can make a real nuisance of themselves by, for instance, tearing at the manger at feeding time or developing stable vices. Besides the tongue-swallowing and lip-smacking chiefly found amongst Trotters, which probably originate in bits being too sharp and tying up the tongue, the most unpleasant of these vices is wind-sucking, with the grunting noise which is emitted when the wind is let out. With some horses this 'bad habit' is only occasionally practised, while with others it becomes an absolute craze, which is practised for hours and causes digestive disturbances, that in turn result in bad condition and sometimes in a fatal colic. I put the phrase 'bad habit' in quotation marks because, like weaving, it is a substitution activity provoked by people, which is done by especially lively and physically active animals, such as those used for competitive sports, which try to compensate for the unbearably long hours standing and waiting in a stable.

## DRINKING

At the beginning of this book we discussed how much the area of activity of wild Equidae depends upon the number of watering places available in a district, since almost all Equidae have to drink daily and only the members of the *Equus hemionus* family, which can survive in the desert, water every other day. By eating succulents that store water in their stems, leaves or roots, some wild Equidae can reduce their need for drinking-water, although water holes are still necessary.

As far as domestic horses are concerned, the amount of drinking-water required depends a little on the food, since it increases when hay is chiefly fed and, logically, at pasture in spring, when there is a larger amount of fresh grass, the quantity of water needed decreases. Added to this, the type of work done by the animal in question affects the extent of his thirst, since horses which daily do fast work and sweat profusely are a great deal more thirsty than those which only laze through the day. The time of year and the weather also play a part, as animals are affected in the same way as we are. Because of self-drinking bowls and water tanks in pastures, to which horses have access day and night, it is difficult to be sure of the amount of water drunk. If they drink from pails, which is usual in training stables and in some of the less-modernized stables, one knows, without thinking too much about the reasons for this instructive fact, that some animals eating the same food require more water than others. I can only explain the great differences in the amount of water needed by individual horses by saying that, as with humans, it is certainly due to a slighter or greater activity of the thyroid gland and the resulting metabolic reaction. There are horses which convert energy into fat and these have a weaker thyroid activity, whilst other horses expend the accumulated energy and are tremendously lively, because the thyroid gland functions much more actively.

In our own stud, where there are no drinking bowls or tanks and where to our own disadvantage we have to carry a fair number of pails, I noticed that at two out of three watering periods a few of my horses took almost no water and then emptied two or three pails in one go. At the beginning I thought this was abnormal physiological behaviour, an individual bad habit found only in half-grown animals and the very old mares which were losing their teeth, since the latter stop drinking until their thirst is greater than their toothache. The postulation of a 'bad habit' naturally does not satisfy a student of behaviour, as it is part of his obligation to explain the reason for the individual behaviour of an animal.

I decided to divide my horses according to the amount and the frequency with which they drank water. It soon became clear that horses with pony characteristics, i.e. animals with very full manes and tails, generally dark brown in colour and of medium size – and here I must explain that with one exception I am writing about horses (animals exceeding 14.2 hh) – all drank approximately the same fairly small amount during each of the thrice-daily watering sessions. A second group presented a different picture: they drank a lot at one watering and contented themselves with a few mouthfuls at the other two times. The third group also drank three times a day but rather more. The drinking habits of all three groups did not vary with the type of fodder supplied or with the outside temperature. This means that their drinking habits remain the same in hot and close weather or when they are fed only on hay. These horses are all extremely active and relatively aggressive, big-framed of a markedly hard and self-sufficient working type, i.e. performance or competitive-type animals.

In veterinary science we speak of coarse-grained horses and of animals with a fine-grained cell structure. Since the fast, slim horse indicates by its cell structure an original homeland which must have been in the dry, hot southern regions, we must suppose that its drinking habit is a relic of wild ancestors, a clear analogy to the drinking habits of wild Equidae today like the kulan, onager and wild ass. All the inhabitants of waterless areas, besides the appropriate characteristics of running animals, have been compelled to develop sensible drinking habits, whereby the daily requirement has to be satisfied in one go. I suppose that in racehorse training establishments the usual practice of giving water in buckets is not only meant to protect the horses from drinking too much water when hot (returning from training gallops, etc.), but this severe rationing of water is also meant to hinder the drinking habits described above, since racehorses in particular belong predominately to the hard, working category of horses.

The lean body of the working racehorse is also partly the result of the rainless habitat of his early wild forebears, which had to cover long distances daily to their water holes. Sometimes in bad years or during the height of the annual drought even these sparse pools and channels were dried up. Most of the animals moved into areas where surface water was still to be found, but certain wild Equidae solved the problem in other ways: those in some of the extremely arid mountain ranges of South-West Africa, the Hartmann mountain zebras, used their forefeet to paw the dried-out river beds with a high ground-water level. Each one made his own water hole about 20 inches deep and between 36 and 60 inches in circumference, and then refreshed himself in the rising water, which was usually much clearer than the brackish standing water found in the natural water holes, which is why the zebras preferred it. Since water is a scarce commodity, the drinking springs that they dig are defended even against close members of the family, and logically here again we encounter the emotion of jealousy, which rarely appears in wild Equidae. The last remaining Przevalski horses in Mongolia are also said to make such water holes and a similar account is given of the Grévy zebras.

Some authors, who principally observe wild Equidae, maintain that the quality of the drinking water is of little importance. This is a blatant contradiction of the widely-held opinion given in books on horse-breeding, that horses are extremely sensitive to poor or strange-smelling water. It does seem that Equidae, if they have the choice, prefer clean, clear, odourless drinking water. Water containing salt or soda is drunk in cases of extreme necessity. Competition at water holes between wild animals, and especially between wild Equidae and the domestic animals of the local nomad peoples, is extremely strong. An extreme instance, according to Russian writers, is seen in the behaviour of the remarkably shy kulans living in the Turkmene Steppes, which condemn themselves to extinction because, although they graze with domestic horses and sheep, they cannot overcome their reluctance to share the communal water hole with these animals even when they are extremely thirsty. Some wild Equidae in zoological enclosures have a shy, inhibited attitude when drinking and this dates back to the necessity for special precaution and vigilance, because in the wild, when they made their daily visit to the vitally important water hole, one animal would stand guard against the omnipresent carnivores and he drank only when the others had finished. In strange surroundings domestic horses often need several days to become accustomed to the taste and smell of the 'new' water, which to us appears to be completely normal.

Horses drink with a sucking action, which is similar to the way we humans drink water from the palm of the hand and not in the way dogs or cats noisily lap up liquid with their tongues. Some horses, when drinking, need to hold their head so that it is in a straight line with their neck, and the modern method of placing water bowls high up is physiologically unwise, especially if small horses then have to drink in an uncomfortable position. I cannot verify the commonly-held opinion that, like some zoo animals, horses actually swallow after several draughts, nor can I verify the pawing action with the forefoot when drinking from puddles or other pools of water. I very much doubt that mud, leaves and weeds could be removed by this means, since water is made muddier by pawing and, apart from this, horses can easily suck water from between leaves or hay; with a gentle movement of the lips sideways, they can easily make a place free of stalks. I also cannot agree with the assumption that 'water games' occur because of excess water in domestication, when horses swill around loudly with lips and tongue in a half empty bucket of water and then often allow the cold water to drip out of their mouths. I think it more likely that it is done initially to rinse the teeth of remains of food and then later extends into a game.

The unique drinking behaviour of some horses which obviously originated in warm regions is the opposite to the drinking habits of many of our ponies. The homeland of most of the pony breeds is chiefly in central and northern European countries, regions where water is plentiful. It is, therefore, quite natural that all the animals which inhabit these regions have not developed such a specific drinking pattern as their cousins from southern regions, but, on every occasion where water was available, they drank in quantity, thus avoiding long journeys to special water holes. One of my stabled horses with a lot of pony blood behaves similarly and the Fjord ponies on free range have also confirmed that, in spite of midsummer temperatures and the many springs and brooks in their locality, they have no special drinking habits. The only exception occurred when the two herds on their way to their afternoon pasture passed a glass-clear reservoir and most of the animals were unable to resist taking a small drink. The rest of the watering periods were so unremarkable that they scarcely had a place in my behaviour records.

Although water is usually available in our locality, when there is deep snow or a hard frost it can be difficult for horses outside to find drinking water. On such occasions they either eat snow or have to use their forefeet to break the ice on ponds and pools, which, strangely enough, zebras in zoological gardens also try to do. My namesake, E. Schäfer, the well-known Tibetan traveller, described something similar (one could almost call it evidence of public-spirited behaviour) in the kiangs, members of the *Equus hemionus* family, which inhabit the high cold steppes and which unanimously, in a markedly rhythmic beat, kicked the ice-cover in order to get at the water.

Foals have to learn how to drink water, because the art of drinking milk is completely different. At first they are quite clumsy, putting their mouths too deep into the water or even biting at it, as they learnt to do with grass or green forage, which they assimilate earlier than water. After several attempts, they are able to keep their noses above water and to use the corner of their mouth just as the adults do, and a proper sucking cleft in the mouth results in a respectable draught being taken.

## PASSING DUNG AND URINE

At the beginning of the book we discussed how horses in freedom use an area similar to our own living arrangements, with special grazing grounds, and rubbing and rolling places, and separate sleeping quarters. As in all very large houses, there are on all the ranges of free Equidae a number of places which serve as a form of toilet, since Equidae are not like other herbivores which leave their dung anywhere. Such places are often found on the paths or at the intersections which the animals frequently pass and which they carefully smell in order to find out whether, and if so when, an acquaintance or member of the family came by, and then they themselves dung on the same spot. Passing dung has an almost infectious effect and the rest of the herd will then add to the heap, so that in time, depending on the number of horses in the herd, an area of several square yards will be covered. On the normal fenced paddocks special dung places are also made, but one gets the impression that they are begun by accident. As horses avoid such places, in a very short time they become surrounded by high grass, and, with ever-increasing amounts of dung, the area soon becomes larger at a rate which depends on the sex of the grazing horses. In the stallion paddock such areas are slower to build up, because stallions dung more carefully. They only do so after close inspection and smelling the dung already there, especially if the dung is that of a mare. An exaggerated *flehm*\* follows, and then an accurate dropping of dung. Mares also smell strong dung drop-

---

\* Translator's note: *Flehm*: there is no comparable word in English except perhaps 'sniff' – when a horse moves his upper lips and appears to 'laugh'. In fact, the horse has got wind of a scent and the action may be due to pleasure or disgust.

pings and sometimes *flehm*, but they do so in a way that is less intense or interested and often dung during their inspection; thus droppings fall a little way from the original ones and the dung place is enlarged by as much as a horse's length. All well-organized studs remove the dung daily to avoid spreading and subsequent infection by parasites.

In the box or in very small yards, special dunging places are of necessity not feasible, but there are many clean horses, which use a certain corner or side in boxes or even in stalls. Some animals which are tied show their intention to move towards their dunging place by walking as far forward as they can and then, when they step backwards, they succeed in making the whole floor dirty.

Both sexes dung in the same way, with their heads lowered, their ears turned back and tails raised to varying degrees. Generally any other occupation such as grazing is interrupted, although horses are perfectly able to dung at the walk or trot if the rider or coachman will not allow them to stop. The amount of dung is primarily regulated by the amount and quality of the food eaten, but it also depends upon breed, individual temperament and the temporary frame of mind. Nervous and frightened horses will dung the moment something upsets them and sometimes pass loose dung in which the undigested oats can be seen – this is especially noticeable in horses just before a race. When a horse drinks water, this also provides a physical stimulation to dung.

In the same way, certain places are preferred on which to urinate or stale, but these are not recognizable until coarse nitrogenous grass has grown on the spot. Staling occurs several times daily and depends chiefly upon the amount of water taken. However, it also depends upon the kind and duration of exercise which the horse has had, i.e. they stale less if they have sweated profusely. When they want to stale, both stallions and mares adopt a similar position: the croup is lowered, the tail is raised and the hindlegs are stretched apart. They wear a concentrated expression on their faces and the ears indicate that this occupation is in progress. No horses like wetting themselves by staling on hard ground and they will retain urine for a considerable time until they find soft ground or straw, both of which encourage them to stale. Stable management economy in which the straw is removed during the daytime amounts to cruelty – and the same applies to harness horses working on surfaced roads – they must be given opportunities to stale comfortably. A stallion grazing with his mares pays attention every time a mare dungs and then he stales on the same spot after he has smelt it. He will then generally *flehm*. Sometimes he is sexually excited and will try to cover the mare concerned. Both mares and foals *flehm* after they have smelt strange, or even their own, droppings, but it is not as marked and does not last as long as with the stallion. In addition, after staling, mares show the so-called 'flash', when they ejaculate the last drop of urine. This action is necessary to prevent vaginal catarrh, which could cause infertility. During heat, mares 'flash' or 'wink' in the same manner, though not necessarily after staling, and often only oestrous slime is ejected.

## REST BEHAVIOUR

Apart from feeding, resting occupies the longest time in the natural life of a horse. Since horses rest in a different manner to many other domestic animals and humans, I should like to characterize their various intensive resting periods purely linguistically with the expressions 'dozing', 'slumbering' and 'deep sleeping'. These three expressions are the same for people and horses, and only the frequency and duration of the resting habits of Equidae varies.

A favourite occupation, or rather non-occupation, is dozing, which all of us have noticed. When horses doze, they have a typical dozing expression on their faces, their neck is held horizontally and is completely relaxed, and, with the weight on the forehand and with a lowered croup, they bend the hock and rest the toe of the hoof. This position was frequently exhibited by draught horses which were pausing or by waiting cab horses. Today many saddle horses doze like this in their boxes or when tied up to horse boxes, as well as when they are turned out. Perhaps some people have the idea that horses can rest fully whilst standing. This is correct only to the extent that, because of the special anatomical structure of their fore extremities, Equidae can relax in a standing position better than other animals. A complicated construction of tendons and ligaments enables them to keep their joints in position without actively exerting the muscles, which results in a real relaxation of the forehand. The situation is different as regards the hindquarters, which also have many sinewy and, therefore, tireless parts, but here muscles are necessary to carry the weight of the body and, like all muscles after exertion, they have to have relief. With fully-grown horses dozing takes up a large proportion of the daily rest period. Above all, during the hot midday hours they enjoy dozing under a shady tree as relaxation between grazing periods. When the weather is damp, cold and windy, they are averse to lying down and stand throughout the night with their backs to the wind, using their tails as protection against both wind and rain. When dozing, horses are in a state of almost complete abstraction, but they can react like lightning if something happens and, depending on how they are standing, they can run away or turn and kick. This retention of the ability to react rapidly makes it possible for horses to doze in unquiet surroundings or in stables with frequent visitors, or even in the middle of noisy street traffic, but, since they often have to change weight from one hind foot to the other, this form of dozing does not replace sleep.

Since adult horses can rest tolerably well when standing, they lie down only when they feel absolutely safe. Apart from their watering holes, wild Equidae are open to surprise by their enemies when they are asleep; therefore in every group an adult animal will act as a guard. This duty is not allotted to any particular horse according to rule, but they change freely amongst themselves and, if by chance all the animals in the family are resting, the last one standing must keep guard until one of those lying down gets up. However, one must not imagine the guard to be a very active sentry who circles the herd looking for trouble. Whichever animal does guard duty (mare or stallion) stands in a dozing attitude near his or her sleeping family, nose towards the wind in order to smell the enemy in good time. It is interesting to note that taking guard duty by turns is not confined to horses in the wild, where it is of course sensible, but is also seen when a number of horses are stabled. Even after several thousand years of

*Illustrations 32-39*

Plate 13  *above:* In the mating season obtrusive colts are forced to keep their distance, whilst the harmless four-year-old (right in the picture) is permitted to stay.
*below:* Completely free-roaming pony mares from northern Spain are kept together by the stallion and driven in another direction.

Plate 14  *above:* Pony stallion from northern Spain drives an in-season mare, with his head low, a threatening facial expression and his tail carried lightly.
*below:* The goal is reached – the pony mare is covered and takes a step forward, while the stallion clenches his teeth on her crest. In front, the two-year-old and yearling are interested spectators. The families keep further apart than Fjord ponies do.

(Continued page 55)

domestication, horses evidently do not consider the most luxurious stable safe enough to do without a herd member to stand guard. It is not certain how the guarding instinct is aroused. I would imagine though that the last animal to remain standing instinctively feels prevented from lying down when it sees its companions asleep.

After dozing, the next most intensive type of resting is slumber. In order to slumber, which is really only a light sleep, the horse has to lie down. This is a much more difficult operation for him than for a carnivore with its supple backbone. To lie down the horse draws his extremities under his body, bends them together slowly until his acutely bent legs, whose muscles have already begun to twitch from strain, can no longer carry his weight and thereby the animal lowers himself on his fore pasterns and rolls on his side. Awkward, old or heavily in-foal mares sometimes let themselves down with a flop. The effort of lying down is such a strain that horses with a leg injury, rather than put any further strain on their injured foot or leg, often prefer to forgo a real rest by lying down, since there is no way they can do so without using all four legs. In order to get up, the horse puts both forelegs forward, with the hindlegs under his body, so that, with a powerful shove of his hind quarters forwards and upwards, an injured forehand only has to be used for a moment to help balance the upward swing.

Plate 15 *above*: Gentle coat-nibbling before mating. In the background Asko looks on with interest.
*below*: Masturbating stallion.

Plate 16 *above*: Haflinger stallion Medicus smells a mare which is on heat. The foal wears a submissive expression.
*below*: Hope is dashed; Ossi kicks out energetically.

In order to slumber, the horse lies in a 'cud-chewing' position like cattle do, his legs under his body and the head held either free or with the muzzle on the ground. Although this sleep is superficial, it is still a real sleep. Even so, if disturbed, the animals can be active fairly quickly, as they merely have to put their forelegs forward in order to stand up immediately. It can be observed that dozing animals remove flies by flicking their tails, twitching their body and nodding their head. These activities cease completely when they are slumbering, except perhaps for an occasional languid movement of the tail. Some horses never sleep in a prone position, because their surroundings are never completely quiet or because they suffer from an inner uneasiness, as seen in neurotic animals. Heavily in-foal mares always sleep in a 'cud-chewing' position as described above.

In a deep sleep, when horses, too, are so sound asleep that they no longer have any sense of perception, they will lie stretched out on their side – head, neck and body are fully relaxed on the ground, one foreleg is usually slightly bent, whilst both hindlegs are distended, requiring a fairly large space. Most horse-lovers will have witnessed this deep sleep in little foals, which, at any time of day or night, will lie down unconcerned, whilst fully-grown horses will only sleep soundly with someone around when that person is regarded as a trusted member of their own herd community.

When in a deep sleep, horses breathe evenly and audibly; with some the single breath ends in a slight groan and sigh, which just before awaking can increase almost alarmingly. Horses also dream when they are sleeping soundly, a fact which has hitherto been doubted, because observers were never sufficiently integrated within a herd to be in a position to give evidence. In much the same way that dogs bark in their sleep, growl or twitch their legs, so that one can almost guess what their dreams are about, horses also express themselves by gentle whinnies with audibly different meanings and by unmistakeable leg movements. I witnessed a particularly impressive example of a clear childhood memory in the dream of a yearling, which, whilst sound asleep, emitted the quite plain whinny of a suckling foal, so that a distant mare, which was about to give birth (for this reason I was sleeping in a box nearby), answered the still sleeping colt with the suitable whinny of recognition.

In order to rest completely, horses need plenty of room for their outstretched legs. As this is always available in their natural sleeping quarters, stalls or small boxes, especially for large horses, can completely prevent them from sleeping soundly. Horses that do not lie down, perhaps because of an injured leg, are never properly rested. It becomes apparent that they need to do so at times when their stable companions are sound asleep, when they too drop off and nearly topple over, at which they wake sufficiently, just in time, to keep their balance.

In contrast to their quick reaction to events when dozing and slumbering, horses awake from a sound sleep like people and regain consciousness gradually. If one touches them, the breathing changes, as I have mentioned, and is unnaturally loud, although the animals remain lying quite motionless. After a fairly long time the ears begin to play, the horses open their eyes, a few seconds later they lift their head and the very heavy breathing ceases. When their deep sleep is over, unlike most people who enjoy a period of superficial sleep, horses get up immediately, stretch themselves by arching their neck with the head at an acute angle, extending the forelegs well forward and stretching the back. If food is available, they begin to feed immediately. Often the hindlegs are stretched out backwards alternately. When they do this, stalled horses like to be able to put a forefoot on a low trough, giving themselves a better chance to stretch fully.

People sleep for several hours without interruption. Unlike us, horses rest in short intervals, and the individual periods of superficial slumber and the single periods of deep sleep rarely exceed a full hour, and normally take place with older foals and all adult animals between midnight and dawn. They alternate rest with short periods doing different occupations like caring for their skin, feeding, staling or dunging, so that, after each span of activity, there follows another rest period, which is often spent dozing upright. Although one rarely sees adult horses slumbering in the daytime, they usually doze for an hour or two after the early-morning and midday feeds as long as this natural daily rhythm is not broken by work of one sort or another. The chief feeding times are also often interrupted by each horse for a break varying from a few minutes to half an hour. The total resting period of adult Equidae – dozing, slumbering and deep sleep – is generally given as about seven hours, but from my own observation I would say it was longer. In hot weather, when the dozing period extends to the detriment of other, more active occupations, it can amount to almost nine hours.

Whilst older horses spend most of their resting periods by dozing standing up, the animals which are not fully grown spend a longer time in the 'cud-chewing' position. This frequent rest lying down appears to be a typical habit of youth and, above all, a childish reaction, because quite remarkably young mares, in-foal for the first time, behave differently to those of the same age which have not been covered. Amongst the herd of Fjord ponies living free during the hot, summer, midday period (apart from foals and yearlings of both sexes) only the two-year-old colts slumber often at the feet of their dozing elders, but not the two-year-old mares which have already been covered by the stallions that run with them all the year round. In my own stables, warm-blooded mares which are not destined to breed do not give up their midday slumber period, whereas their young three-year-old contemporaries behave like the older mares and no longer lie down. Such small observations help with the positive or negative results of a pregnancy test, but this can only apply to horses which have sufficient trust in people not to allow their normal reactions to be influenced by their presence.

Apart from suckling its mother the life of a very young foal is dominated, like a human baby, by the necessity to sleep. Above all, very young foals react in a spontaneous and lively manner similar to that of human children: after a short period of exertion in which they execute wild buck-jumps or gallop as fast as they can in circles round the mare, they are suddenly overcome by such extreme tiredness that they almost throw themselves down and fall fast asleep. One can see them at any time of the day or night in a variety of places, which need only to be dry, lying like dogs with their head between their forelegs or stretched out obviously in a deep sleep. Their mother grazes or dozes nearby, since at the beginning the mare is never far away from her foal, which still needs protection. Later on the mother's love is not so intense; sometimes, whilst grazing, the mare follows the herd and gets so far away that, when the foal wakes up and finds itself alone, it is almost frightened to death. It jumps up, neighing loudly, and gallops as fast as it can after its dam, which by this time has answered the foal's cry for help; it then generally comforts itself by suckling. The duration of a foal's resting periods greatly exceeds that of full-grown horses and, of course, the time decreases as the foal grows, and begins to graze and to feed itself.

Young foals seldom doze. If a suckling foal stands and dozes for a long time, I take this as a serious warning sign, because usually the animal does not feel well and, more often than not, it even indicates the beginning of a serious illness. The startling saying from cowboy films that 'a man won't die as long as he stands on his own two feet' appears to be even more applicable to horses, because they obviously feel they have to remain standing as soon as they feel ill as they fear that they would not be able to get up and flee. This is particularly the case with sick foals and very old horses.

In the description of the range, rhythm, and pattern of movement of Equidae, it has already been seen that horses do not lie down just anywhere but look for definite sleeping quarters. A human being would find the sheltered and screened sites of their area the most comfortable, but, if the choice is theirs, horses prefer spaces open to the wind in all directions. These exposed windy sites provide horses living in the wild with a much stronger feeling of safety against approaching enemies than perhaps a sheltered wood. Such places are essential to a horse's feeling of comfort when lying down – an instinct which even our cherished domestic horses have retained. Even after such a long period of domestication, they feel happiest in the same places as their wild cousins, places which are quite the opposite to our ideas of comfort. Horses that are kept out usually use field shelters only to avoid flies; they lie down in the open to sleep. The regret expressed in a racing journal which I read recently, that racehorses in the warmer regions in the U.S.A. were housed in open, airy boxes of coarse wire-netting, through which they could be seen, and 'therefore were never really able to rest', is irrelevant since, apart from the physical well-being of these animals in surroundings which were not so stuffy, their psychological comfort must have been greater than that of many European racehorses confined in their prisons.

Aside from the essential feeling of safety, horses require their sleeping quarters to be on dry ground. For this reason wild zebras prefer the so-called scrub steppe and some domesticated horses create their own 'scrub', when they choose a high position, where the wind blows, and then graze the grass to the roots, so that they can sleep on the dusty ground. The decisive factor of a good bed in the stable is not so much the softness of the straw but its dryness. Therefore, the idea of providing special sleeping quarters in the box by putting down sand, impractical though it would be as a natural method of horse-keeping, is not so peculiar. Hardy, robust horses will lie down on powdered snow and, even in winter, they still choose places where the wind blows and their senses are not impaired, rather than sheltered valleys.

The customary distance between individual animals remains the same even during sleep: the antisocial horse prefers to rest alone a few yards away from the others, while grazing companions and playmates lie very close together, sometimes touching one another.

For prolonged dozing as a group there are also special places, but these do not need to be so open or accessible – not so 'safe' – as the dry, airy quarters for slumber and deep sleep periods, because most of the herd rest in a standing position. In summer, shady trees are sufficient or field shelters, etc., and for the remaining short daytime breaks no special place to doze is needed.

## SOLITARY GROOMING

Horses spend a considerable part of the day caring for or grooming their skin. This is absolutely necessary for their well-being and can thus be described as behaviour conducive to comfort. The skin is not just a protective covering which encloses the body; it is rather an active organ with many important functions, indispensable as the seat of the sense of touch and feeling of heat or cold.

Since other writers have gone fully into the matter, I shall only touch on the connection between skin activation and nervous stimulation, for example sweating from fear or pain. The connection is basically explained by the fact that both derive from the same cell tissue formation. All we need to do here is simply to establish that horses require a completely serviceable, healthy coat, and, as wild Equidae are not groomed by humans, they develop a fairly complicated and above all a truly characteristic and marked behaviour conducive to their comfort. This makes it possible for them to groom and keep their skin and coat in condition themselves. Although this equine behaviour of grooming the skin also has to fulfil a social function, which we should not underrate, I will deal first with individual care of the skin and describe the things which an individual animal does for his own well-being.

Rolling is the most obvious individual action in caring for the skin. To the regret of their owners, even well-groomed horses find it absolutely necessary to make full use of an opportunity to roll; the urge is sometimes caused by grooming, which completely nullifies the trouble taken to see that the horse looks well-groomed. Although horses will roll almost anywhere where the ground is not too hard, they usually prefer a special place and above all somewhere that is dusty or muddy. Wild Equidae allow themselves a dusty bath in the hot midday hours, when the dust is at its driest, so that afterwards they can carefully shake out their thickly-powdered coats. So too, where horses run free, these dust bath places can be found, whilst most stabled horses, when turned out in the paddock, find a dry place somewhere. Horses sometimes make such a place for themselves when they eat the grass to the roots and then, by frequent usage, eventually have a flat hole for a dust bath.

The reason for the need of all wild Equidae and horses that are kept outside to give themselves a daily dust bath may be the fact that fatty sweat, night-time dew or rain matt the hair together, so that their coats no longer lie naturally on their bodies, and they try to loosen the hair to get it into its proper position. When stabled horses are groomed, the effect is the same. Little bits of skin are loosened with the dandy brush, body brush and cloth, and thus produce an unnaturally smooth, shining layer which is stuck together.

In order to roll, horses lie down in the normal way, having just inspected the spot. Their intention is clear to the casual observer by their quite remarkable conduct when they put their head down, hold their tail high and straight, prick their ears, and inspect and smell the chosen place. Usually they circle it with small steps and even paw the ground before using it. Once down, with head and neck pressed on the ground, they give the face and cheeks, which they sometimes actually rub into the earth, the same cleaning treatment. After working thoroughly on one side of the body, some horses swing themselves over onto the other side – really supple animals may change over several times. Some horses position themselves in the 'cud-chewing' position and then paw the ground with one forefoot, before once more having a good roll. Horses which cannot or will not roll right over, such as heavily in-foal mares, lie down twice and roll first on the one side and then on the other. Since this exercise seems to be catching, one rolling horse soon encourages others to join in this pleasurable occupation.

Rolling is in fact such an elementary necessity for many horses that they will try to do it in unsuitable close quarters and in narrow stalls or, with clumsy horses, even in large boxes. The result is that very often the animal becomes cast. When rolling over to the other side, they get themselves too close to the wall and their acutely bent legs have no power to get them back to the correct side again, or the neck is too bent and they then urgently need help to enable them to push themselves back into the right position for the forehand to have room to stretch. When a number of horses are tied, irreparable damage, such as nervous paralysis, can result if a horse remains long in such a cramped position, which is why, for example, circuses always have a night watchman. A bed of clean straw is an encouragement to roll, as it is a substitute for a natural dust bath, and most stabled horses give in to this urge without hesitation. I think it is quite irresponsible, if horses have been bedded down on clean straw or sawdust chips, to leave the stables until all horses have rolled and got safely to their feet again.

Up to 1½ years of age youngsters seem to need to roll less than adult animals, perhaps because up till then they spend the greater part of the day lying down on their side and can thus rest and clean their sides. All young horses have to learn to roll and they seem to get very excited when for the first time they see their mother doing it. When the brood mare rolls for the first time after having the foal, which is usually not for some while, since mares do not trust themselves to roll around in a box because of the foal and even in the open they are so careful of their new-born baby that they forgo their own comforts, the foal stands and looks on with a surprised, even puzzled, expression and sometimes gets its mother's legs on its head. This results in the little one also wanting to show off its energy and, rather like a rocking horse, it will jump wildly around its dam, producing the full range of behaviour mannerisms in order to impress, with a sharply bent neck and tail straight out. I had an amusing experience of this kind with a four-month-old, very energetic colt foal, who watched an adult mare rolling at first with great interest and then suddenly threw himself, as it were, into the middle of the turmoil. Whilst the dam was rolling from one side to the other, the colt landed right on the stomach of a horse that was a stranger to him!

Besides the dust-bath, after which all horses shake themselves vigorously, some horses want a mud bath and, if there is no real mud puddle around, they look for the softest, earthiest place on the meadow, in which they roll with great pleasure until they are covered in mud from head to foot. This behaviour, which is distressing to 'tidy-minded' horse-owners, was originally necessary to protect the body, because dried mud is a screen against troublesome insects. Later, when the mud crust naturally falls off, it removes dead skin and hair, an effect which is nowadays replaced by rough curry-combing. Horses enjoy rolling in the snow and strangely enough so do other Equidae, like zebras, which never come in contact with snow in their native environment.

Since rolling seems to give horses such enormous pleasure, one should give them the opportunity to do so as often as possible. The hysterical cries of many horse-owners, when their favourites, by way of a treat, are let out and immediately take advantage of the opportunity 'ungratefully' to dirty their carefully-brushed coats is typical, as I have often said, of the selfish behaviour of so-called horse-lovers. In some American training establishments, in spite of the fairly rigorous training methods, more attention is paid to the psyche of the horse than is common in most European countries. Here, when the horses have finished work, they are allowed at least on the lunge to roll to their heart's content – and, in spite of staff shortages, it is an example which could be copied.

Almost all horses like going into water and by instinct can

swim quite well; for this reason it was a customary kindness to allow the hard-working farm horses to have an evening bath in the horse pond. Some horses even roll in shallow water, which possibly indicates that pawing before drinking in small pools, which has already been referred to, may signal an intention to roll which is not actually carried out.

Equidae also possess a further repertoire of behaviour connected with skin care. For example, they shake themselves vigorously not only immediately after a dust bath, but also when their coat has become sodden by rain or snow and after bathing. They begin shaking with head and neck, followed by the whole body as far as the dock. To do this they adopt a stance rather like a sawing-trestle, so that they are properly balanced to shake themselves. Frequently only the head will be taken from side to side – completely the opposite to the vertical head-shaking to keep flies away. Insects can also be removed by moving the skin muscles and twitching, by a carefully-aimed blow from the head with the mouth closed, by stamping the legs and, above all, by constant use of the tail. As long hair can sweep over a considerable part of the body, the practice in many countries a few decades ago of docking the tail was dreadfully cruel, since the wretched horses lost their only effective defence against flies. In wet areas where they can breed easily, insects are a menace and the various shapes of the tails of different species of Equidae (from the short tuft/tassel of the donkey and rather hairy brush-like tail of the steppe zebras to the thick tails, which almost touch the ground, of many of the pony breeds) seem to bear a direct relationship to their original environment. Amongst our domestic horses the Arab, with his thin, silky, rather scanty tail compared to the bushy hair of a Norwegian pony, is probably the most obvious adaptation to an insect-free environment. His extraordinary sensitivity to these little pests, which is shown by his nervous behaviour in hot, close weather, similarly points to an environment free of pools or other stagnant water.

Nibbling, scratching and rubbing are other exercises that serve first and foremost to care for the skin and rather less to keep away parasites, unless, of course, the coat has lice or mites. Horses chiefly nibble their flanks, parts of the croup, belly and especially the legs; all parts of the body which are difficult to reach are rubbed against a hard object when they irritate. Scratching is done only with the toe of a hindfoot and the parts thus within reach are behind the ears, the ears themselves and parts of the head and neck. It is a short operation, and always very carefully and thoughtfully carried out. Sometimes this irritation, especially around the poll, is caused by constantly wearing a head-collar. When they are tied and get a foot caught in the rope or head-collar, this results in panic and horses can do themselves great damage.

All Equidae enjoy rubbing themselves frequently and for lengthy periods against a firm wall or post, and it has been reported that the African zebras carefully look for ant hills with very rough surfaces, which in the course of time they polish smooth. If no firm object is available then Equidae will rub against one another. The domestic horses prefer trees and paddock posts, and can, thereby, demolish strong fences. Some owners put a scratching post in the centre of the paddock, which the horses enjoy using, but perhaps a much better idea would be to build an artificial ant hill like the one which the director, Hediger, has built for the zebras in Basle Zoo.

Horses will rub every possible place on their bodies: their face, the under parts of their neck, their crest and their dock. They appear to enjoy intensive rubbing, as is easily seen by the expression on their face. From time immemorial humans have rubbed horses in order to obtain their friendship. This is especially useful in obtaining the trust of foals, and to rub them on the crest or croup is better than clumsy patting, which is often understood by the foal as aggression. Rubbing is a means of communication and, by using it as such, we are entering the realm of genuine social behaviour.

# 3

## SOCIAL BEHAVIOUR

In the animal kingdom there are species which are to be found only singly or at the most in pairs, and others which are very sociable and have a society with a clear social structure. Solitary and gregarious species may be closely related, for instance the solitary fox and the pack wolf, both of which belong to the dog family. Many animals only associate in pairs in order to mate or to raise a family.

Equidae are amongst the gregarious category of animals which formed a pattern of social behaviour that varied slightly from species to species according to their environment or their degree of physical development. It is thought that 70 million years ago the ancestors of all Equidae, when they were still polydactyl creatures creeping around in the undergrowth of tropical jungles, lived singly or at the most in pairs and that their behaviour resembled that of the tapir, which is one of the closest relatives of the horse and which, even today, apart from the mating season, has remained a solitary animal. While the social life of the tapir is confined, and therefore correspondingly inadequate, to mating and mother-child behaviour, those animals which stay together in a family period or which form a pack or in extreme cases a herd have developed a mode of living together which guarantees, as far as possible, an association free from friction. This mode of living together in numbers, with a whole range of permitted and not permitted conduct, is called social behaviour.

The natural shades of social behaviour in all gregarious Equidae can no longer fully develop in our domestic horses, which are so limited in their freedom of movement, and they are even modified in horses out on grass or in ponies on free-range with a stallion running with them. Social contact is stunted in stabled horses in many ways. The need of our horses to enjoy life to the full in a community is so overwhelming that they are prepared to overlook considerable differences in size, breed, age and sex, and, in extreme cases of being kept on their own, even the differences between species. There have been many stories of the beneficial effects of a substitute animal companion, like a goat or the famous cat belonging to the Hungarian mare Kinzcem. However, these substitutes are really emergency associations, which horses accept for lack of one of their own kind or of a closely related companion like a donkey.

*Illustrations 40-49*

Plate 17   Covering by hand.
    *top:* Ganda, a Holstein mare, squeals softly in the teasing pen and the position of her ears shows her willingness to mate.
    *centre:* Fabulus, the Holstein stallion, is importunate. Ganda holds her tail to the side.
    *bottom:* The Thoroughbred stallion, Markgraf, *flehm*s and Ganda shows by her stance and mating face that she is ready to be covered.

Plate 18   A foal comes into the world.
    *above:* The foal is almost free of the caul.
    *below:* A horse's head always appears small and noble immediately after birth, because the ears lie flat and slowly come upright.

(Continued page 71)

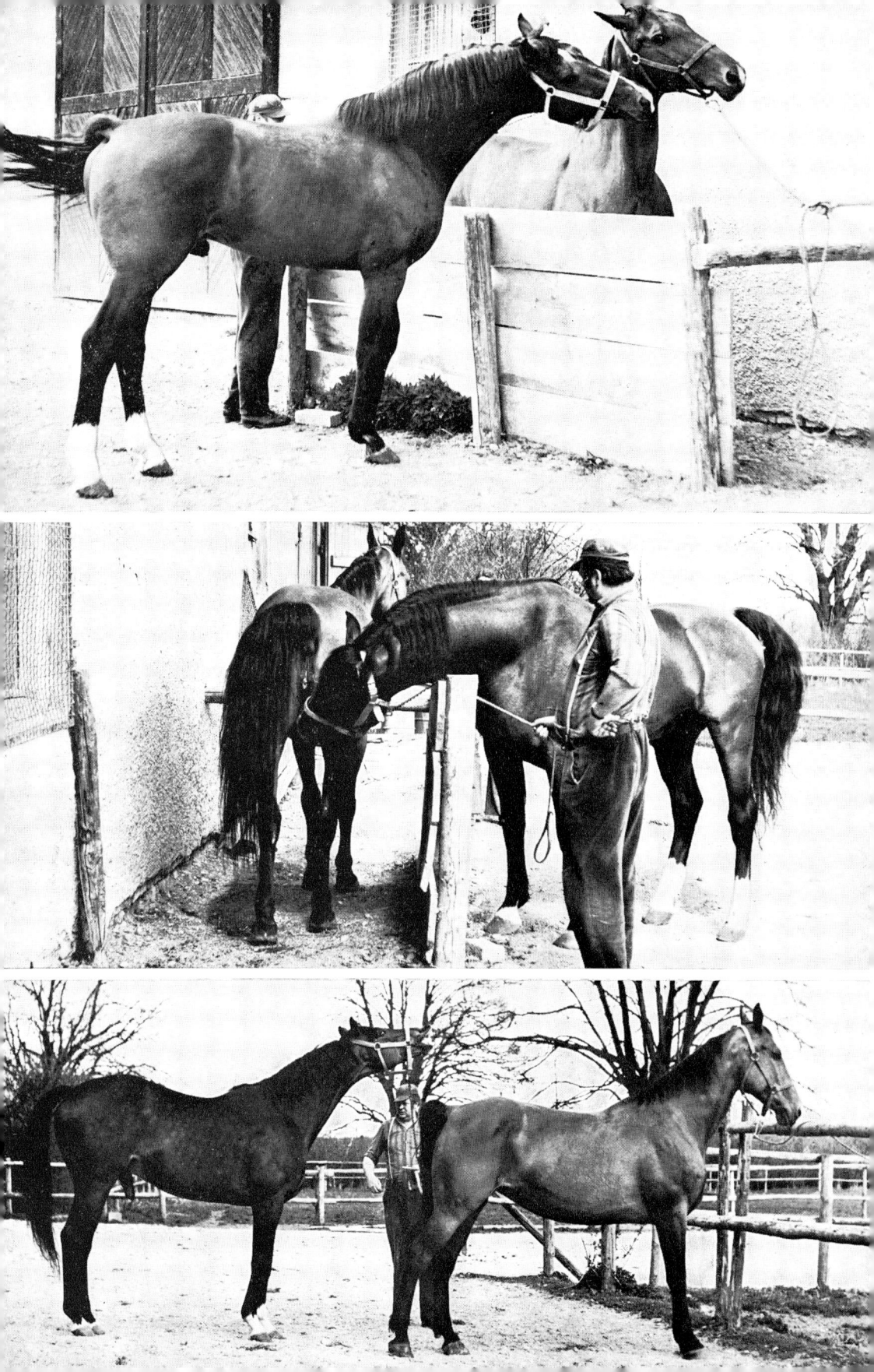

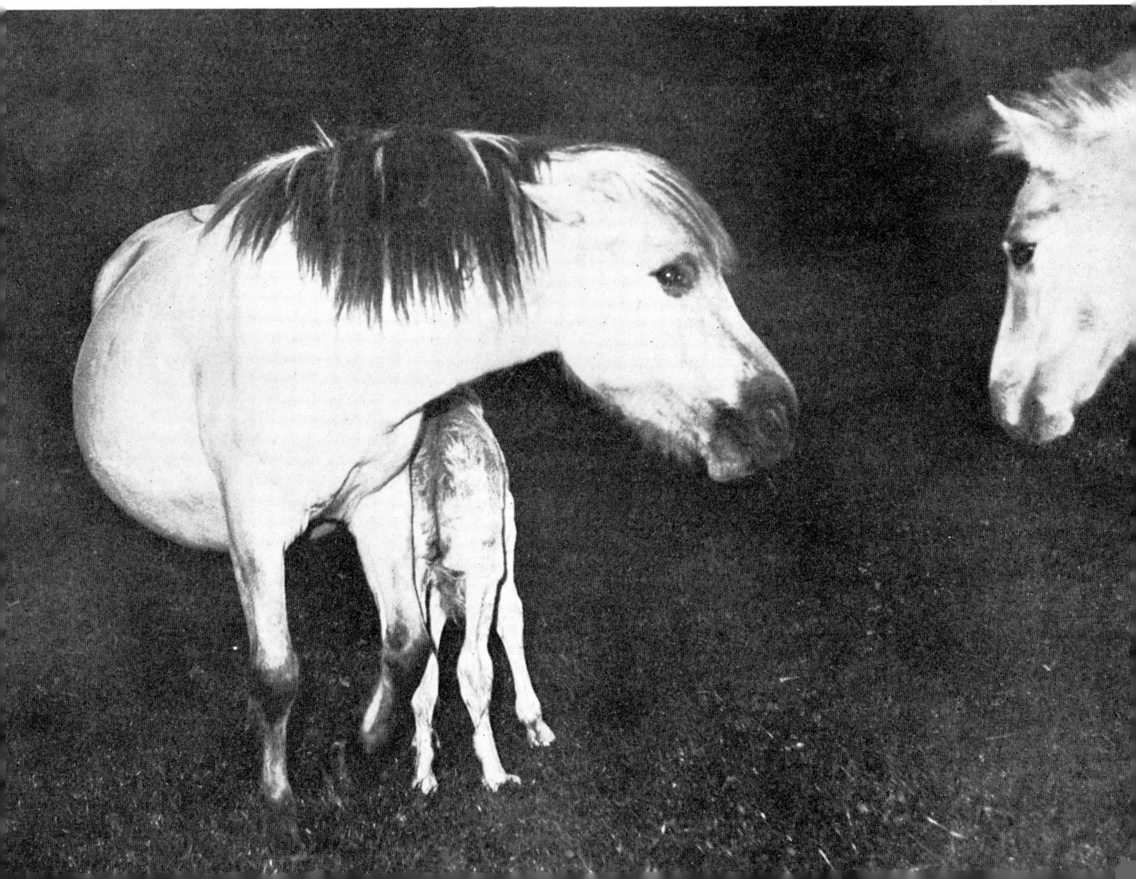

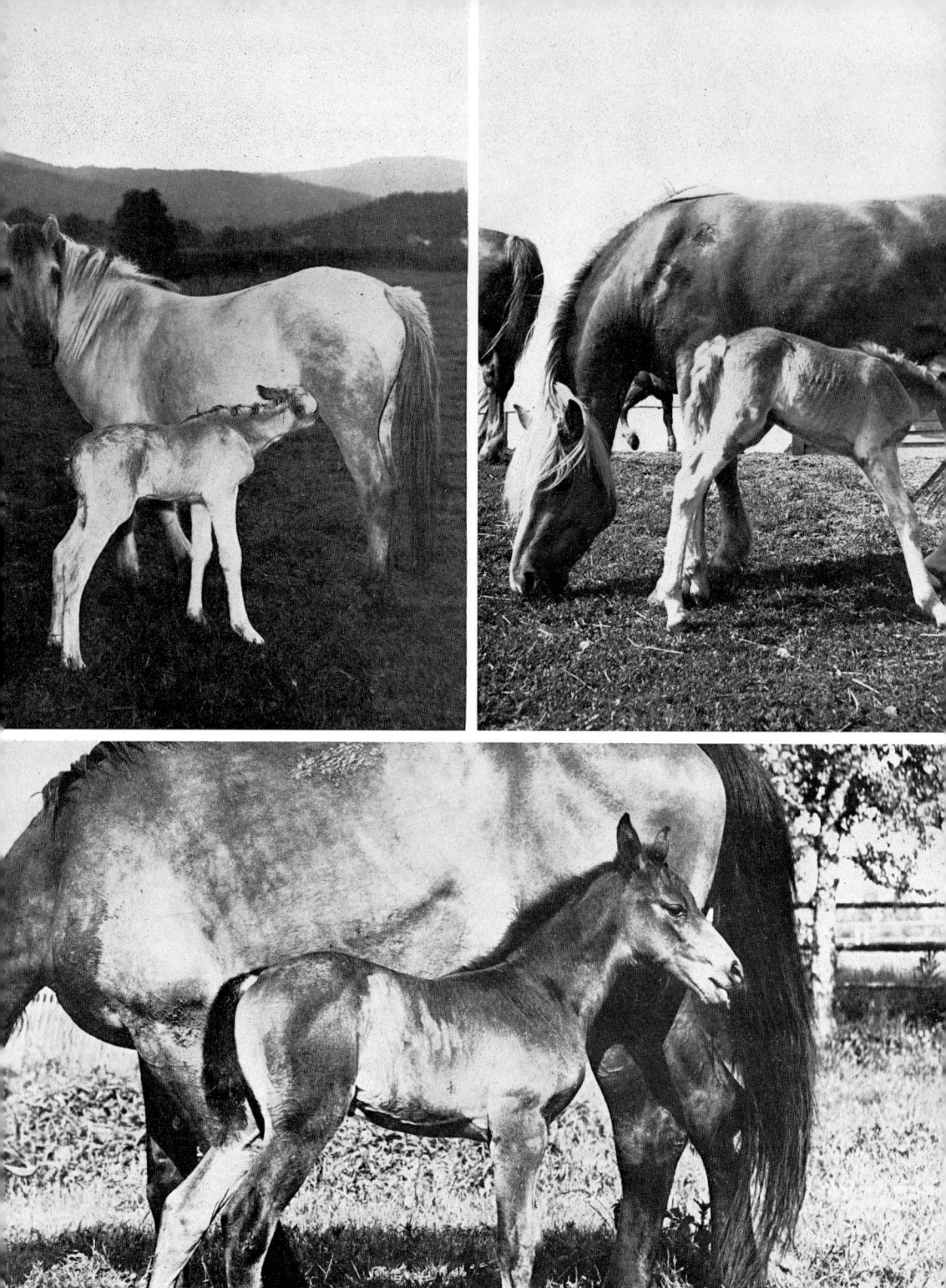

# SOCIAL GROOMING

Apart from mother-child behaviour (which we will leave aside for the moment since it will have a chapter to itself) two horses by themselves will form an association. As regards recognizable, loosely-knit social behaviour, mutual care of the skin is the first sign, when both partners nibble parts of each other's body which cannot be reached alone. This exercise is carried out when two willing volunteers meet at any time during either grazing or dozing periods, and it can last from a few minutes up to half an hour. Animals wanting this kind of contact walk towards each other slightly inclined from the forehand and wearing their 'cleaning' expression to show their intention, to which the prospective partner replies with the same expression or brusquely refuses. If they have agreed, they usually start on

Plate 19  The recognition phase
*above*: A twenty-minute-old warm-blooded foal examines his mother with all his senses, while she is in the throes of after-birth; her look is introspective, the ears are turned towards her labour pains and the clear, hard line of the jaw muscle betrays the strain she is under; the tired expression and the loose lower lip are both signs of fatigue.
*below*: Fjord pony mare with newly-born foal prevents premature contact with other members of the herd by threatening a larger foal. This foal, which a few hours later befriended the newborn, wears an expression of interest.

Plate 20  *above left*: Foals born relatively small have less difficulty in finding the udder. Fjord pony mare Fanta with her foal, which is only two hours old.
*above right*: Very large foals find it much more difficult, as the neck must be twisted. Haflinger mare, Ossi, with her seven-hour-old foal which is over three feet high.
*below*: Aischa, a Holstein foal, shows her grooved, suckling tongue clearly.

the mane and sides of the neck or around the withers, nibbling industriously to pull out loose hair with the teeth – this has a similar effect to when they are groomed with a curry comb. This exercise is carried out with varying intensity and, if the horses are totally absorbed, one can hear the muffled noise of biting teeth. From the mane and withers right along the back, the coat is thoroughly worked over until the animals eventually stand head to tail. Often they turn and work over the other side. Generally this coat scratching is done by two horses, but exceptionally a third may join in for a short time.

If necessary, coat scratching may be done mutually by quite different animals and, for want of a member of its own species, even a person may be invited, by pulling and pinching at the clothes, to take part in this pleasant occupation, but often the horse's intentions are misunderstood by us. If we accept the invitation, we can help the horse to look upon us as a valued companion. It is difficult and sometimes technically painful to fulfil the wishes of a vigorous and boisterous animal – and, when one invites a timid foal to play at coat nibbling in order to gain its trust, one must try after a few minutes to divert its nibbling energies towards its mother, by pushing the little animal in the right position and transferring its nibbling from one's sleeve to her flank. The dam is, in any case, initially the foal's coat-scratching companion and it will soon master the action and technique of grooming.

## FRIENDSHIPS AND ENMITIES

Horses which live more or less natural lives certainly do not scratch the coats of all individuals equally. Since in every herd there exist friendships and enmities, likes and dislikes, which as with people can be quite spontaneous, friends are always preferred as partners. At the beginning, of course, it is without exception the mother; later youngsters of the same age become partners – with young colts, mutual nibbling soon ends in a game of running or fighting. Older animals sometimes have lasting coat-scratching friendships and with mares it is usually their own daughters. During the pre-heat period mutual grooming contact between the stallion and mares serves chiefly as an opening to the mating game and as a prelude to sexual behaviour. Stallions occasionally nibble the skin of their own foals and, when it happens to be a really masculine pasha, this is rather moving, because one does not expect such a charming 'motherly' gesture.

The way in which spontaneous friendships and enmities can occur between horses is seen from the following experiences. One night at about 4 am, a colt foal was born in the herd of the Fjord ponies which I have already mentioned. When the general excitement over the new member of the herd had died down, a four-month-old filly foal remained so persistently interested in the new-born baby, in spite of threats from the young mother, that in the end the mare stopped trying to drive off the little filly and let her stay. Although there were a number of playmates of her own age, during the next few days the filly only left the foal in order to suckle her own dam, which was often grazing quite a long way off. We never discovered if this was a type of maternal instinct, like that of little girls who play with dolls, or if it was simply the beginning of a normal childhood friendship.

Even more spontaneous was the enmity which lasted for two years between a warm-blooded mare and a Trotter mare on my own stud. Without any reason at all the three-year-old Aischa went for a Trotter which had just been introduced to the herd. One rapid attack was made after another; every effort to get to know the other animals was forbidden until the newcomer kept her distance and grazed alone. When at last Karla was accepted by the other members of the herd and could graze at will amongst them, during the days that followed, one could often see how her arch-enemy suddenly remembered that she had to drive off the stranger, whereupon she usually started to look for the lower ranking animal and isolated her with hasty attacks. During the years that followed, the Trotter mare, which began at the bottom of the order of rank, rose to a higher level, without, however, reaching that of the warm-blooded mare. Aischa, even today, never misses an opportunity to kick or bite her enemy.

Friendships between horses usually develop between those of compatible temperament and character and which need the same amount of exercise. Thus four of my colts of the same age paired up from the moment they met: typically the Arab colt attached himself to a very well-bred, quick-witted, warm-blooded colt, whilst the Haflinger pony with its robust character preferred a rather solid, easy-going playmate. This arrangement lasted until the two warm-blooded colts were sold and only then, of necessity, did the remaining partners play together. A knowledgeable breeder is well aware of the importance of these herd friendships to the physical well-being of the horse. In breeding Thoroughbred horses corresponding conclusions are drawn and, wherever possible, uneven numbers, especially in herds of young horses, are avoided. Experience has shown that the 'odd foal out', which cannot pair up with a companion, does not develop so well, although there is nothing wrong with him – this is an obvious example of a spiritual deficiency.

## HIERARCHY

In spite of mutual care of the skin and a possible ensuing friendship there will very soon be an order of precedence in a small association such as arises when two horses are arbitrarily kept together, when one animal is lower than the other as is shown by the fact that it is always last to enter the stable or to drink from the water trough. Through lack of knowledge of the subject, one is always inclined to believe that the natural way of living is perhaps better, happier and, above all, fairer than our form of society with its Establishment, privileged and not so privileged classes. In Nature there is no such thing as equality and, therefore, socio-political ideas which propagate equality of opportunity for everyone are actually unnatural, even though they are well-meant and commendable, as their idealistic background may show. In an animal society there is no equality; on the contrary there exists in all species which live together a formal and strictly observed hierarchy, that is guarded more jealously and consistently than in many human systems of society, however reactionary they may be. This absolute order gives the lowest individuals, which by no means appear oppressed or unhappy, a feeling of security because they have a specific position in the herd, and know exactly what they may and may not do. It is a mistake, and for the sake of peace in a community of horses it is not advisable, to give preferential treatment to horses in the lower ranks, if, for example, with a sentimental idea of justice, one picks out the smallest or the youngest to receive the first tit-bit and thus causes a permanent hierarchical jealousy. A herd of grazing horses remains a peaceful association as long as the hierarchy of each individual is clearly understood amongst them. If strange animals are introduced and left to themselves, battles for precedence will break out at once, but they subside when all the horses have ensured their proper place. Each animal takes on one by one the rest of his fellows, whereupon the strongest personalities, which are not necessarily either the biggest or the strongest animals, but are often simply superior temperamentally or in their ability to react, thus become top of the hierarchy. Sometimes these bitter battles last for several days; it depends exactly how ambitious and bellicose the horses are which are fighting for leadership. The weaker characters will recognize the 'lead' mare\* and follow immediately behind her without resistance, while amongst themselves they will sort out the middle and lower orders of precedence. Some time later we again have the usual picture of a peacefully grazing herd, in which only the occasional fight can be seen. In Nature it is rare for a number of animals which are complete strangers, having differing laws, to be forced to remain together for longer than it takes to escape, so naturally battles for supremacy rarely take place.

Since strange horses always have to establish their position in the hierarchy, a fierce quarrel amongst equal adversaries can scarcely be avoided. Of course, one can ensure that there is a fence between them, as in zoos, where new zebras may be put into a neighbouring enclosure so that the animals can smell, examine and get acquainted with each other. Horses too can be put initially in adjoining paddocks. It is always useful if neighbours grow to like one another, so that later, when one of the animals naturally takes the lead, the hierarchy battle is only symbolic. Unfortunately sometimes the opposite happens: from the start the horses dislike each other and are on the watch the whole time for the chance to meet and teach the other good manners. The superiority of a horse is shown when it can with impunity threaten another horse, take its feeding place and has priority over a special rolling place. All of them jealously guard their status and every slight by a horse of

---

\* Translator's note: Perhaps more commonly known as the bell-mare, because in olden times the pack-trains in Great Britain were led by horses with a bell around their necks – and although the bell is no longer used, in herds of horses the leading mare may be referred to as the bell-mare.

lower rank which does not respond at once to a clear warning of ears hard back and tail threshing from its superior to go away will be punished by kicks or bites. Equality amongst horses does not come into the question. The respect due to a very high-ranking horse is immense. For example, if the lead mare remains standing in the entrance to a favourite paddock, the rest will not dare to pass her, even when they are being driven from behind with a whip. Therefore, if grooms are unaware of this and are careless, such situations can easily result in injuries to the animals, because, when other horses are milling round, the higher ranking horses kick regardless, in order to keep the necessary individual distance. When horses are let out to pasture, the strongest equine personalities frequently take good care to gallop at the head of a herd that has been together a long time; they could easily be overtaken by the younger ones, but the latter risk it only when there is plenty of room. Experienced racehorse owners try to buy yearlings showing energy, courage and strong initiative, which would have probably held a high place in a herd, because such horses generally make very good racehorses.

The social balance of an existing herd is disturbed by a newcomer which, after one or two days of shyness and uncertainty in his strange surroundings, will understandably fight for his former equivalent position, i.e. he will not without dispute allow himself to be placed lower down. In addition to this, young equine followers occasionally demand a readjustment in their positions, and so, too, do mares that have been covered for the first time and are clearly in-foal, because in this case they apparently go up the social scale and have to be allotted a new place amongst the older animals.

The social rank of a horse within the herd community appears to be regulated not only by his environment but, like everything within the psychological sphere, it is dependent upon heredity, because interestingly enough the offspring of high-ranking horses generally achieve a high position, whilst the foals of those horses lower down remain on the whole in the lower ranks. The necessary characteristics of a mare high up in the equine hierarchy such as conspicuous energy or even aggressiveness are inherited in exactly the same way as external characteristics and will be apparent in the foal. Another aspect to be considered, which may be modified by environment, is the similar social classification of the suckling foal and his dam: every member of the herd respects the foal of a high-ranking mare almost as much as they respect the dam and, because of this, he is in general more cheeky than the other foals, since, when danger threatens, he can seek protection in the authority of his mother. These young animals grow up with a much greater feeling of confidence than the foals of weaker mares, for whom, for psychological reasons rather than physical ones, it is extraordinarily difficult to reach a higher rank than the one obtaining at birth. Even in this situation, Nature does not allow the prospect of equality. Since foals which are raised under natural conditions are not weaned at four to six months, as is usual in most studs, but lose their close contact with their mother only when the next foal is born, the strong influence of the mare's rank, the foal's dependence on her position and the mare's protective behaviour last quite a long time.

If a person wants to be tolerated within a herd system, not as a Gnu in a zebra community, but to associate with his animals free of danger, he must without fail make himself one of their highest ranking members. It will depend upon opportunity whether he then, in a figurative sense, fulfils the function of the lead mare or the herd stallion. If the horse-owner or breeder is not in a position to carry through his claim to leadership consistently, which of course cannot be achieved by force, he is left at the mercy of his horses' goodwill and at the slightest emergency he is in danger, since the most harmless reproof by a high-ranking horse can, under certain circumstances, have an unfortunate effect upon a human being.

## THE EQUINE FAMILY

It may have struck the reader that, in the description of hierarchy, the stallion was not included, although in a natural group of Equidae he is always the highest ranking animal; he can subordinate others without contradiction and his sudden appearance is usually sufficient to restore order to warring mares. This behaviour, which is familiar to some people, since the practice of keeping hardy pony mares on a free range with a stallion has become more general, agrees with our current ideas. Possibly we have been influenced by numerous books on horses and, through watching television programmes about horses like Fury or Flicka, we have become accustomed to imagine that a herd of horses consists of a great number of animals which are guarded, led and defended by a magnificent stallion. Up to now it was generally believed that most of the stallions belonging to the species Equidae were absolute overlords, which were constantly preoccupied with increasing the size of their herd and circling their harem to protect them from possible rivals. This belief has been strengthened in those people who are particularly interested and better informed by research, lectures and films on primitive horses like the Dülmen ponies in the Merfelder Bruch and the Camargue horses of southern France, so that our sub-conscious has been impressed with the picture of a large number of mares and foals and, dominating them, a very energetic stallion. No one has taken the trouble to remember that the same number of colt foals (and later adult entires) will be born as filly foals and that in all the primitive herds in every part of the world these yearlings are caught and removed from the herd community or, when in the hands of the various Asiatic nomads, are gelded. This interference is an enormous intrusion on the social system and may cause its complete disorganization. A few writers, in fact, have written about a matriarchal social structure amongst genuine horses, but they confined their observations to certain definite types of horses and did not represent general opinion. Others have assumed, especially amongst the asses, but also amongst other Equidae not belonging to the horse groups, that there is a similar matriarchal hierarchy with a dominating lead mare and a single stallion which, after a great deal of chasing, was able to cover a mare only during the mating season. These beliefs were based chiefly on the behaviour of animals in zoos, supplemented by observations made in the wild – mostly of a scanty nature and very difficult to carry out – built on just as unnatural social systems as were the researches on semi-wild herds with one stallion running with them.

## The family structure

It was only during the last few years that H. Klingel, a German zoologist, was able to make a comprehensive study of the social organization of different herds of zebras in the wild state, which threw a completely new light on the subject. If fresh information causes doubts to arise about existing concepts, it is right to examine critically the accuracy of earlier scientific theories in the light of modern research. By a stroke of luck I had the opportunity to do just this and to study the growth of a genuine natural herd under ideal conditions amongst the Grant zebras in the Serengeti. As I have repeatedly said, even the pony herds kept under fairly natural conditions with a stallion, his mares and their filly foals, represent an unnatural equine community, because they are not complete. In most countries only one stallion is kept in a herd, partly for reasons of stud management, partly because of space, but principally, however, from fear of the possible consequences of the supposed bitter fights between stallions. The Fjord pony herd, which has been frequently cited, also served to help me check the results of my experience in the Serengeti. The interesting thing about this herd of domestic horses was not their free range in a very large area, but the fact that they enjoyed a natural social system, because in the herd of thirty-four, comprising eleven brood mares, two fully-grown family stallions and a number of sons of various ages, the offspring of the two stallions ran with them. In every equine family, representing the smallest social unit, there is a stallion as the head of the family and a mare (probably with a foal and her yearling) to which the other stallions aspire in order to increase the number of their wives. The desire to own a large number of mares complies with the notion held to date. However, it would be impossible for even a strong and active stallion to manage harems even half as large as those given to domestic stallions by some pony breeders. Since, on average, births are equally divided between fillies and colts, there are as many adult entires as fillies, and, because all horses possess the mating instinct, competition in the wild is extremely fierce.

A herd of horses is by no means a homogeneous organization; depending on its total size, it will be divided into few or numerous family units — in one case it was divided into two large family communities managed by one or other of the two Fjord pony stallions. The five-year-old Endo, the stronger and rather more energetic of the two, owned ten of the eleven mares in the spring. His full brother, Findo, younger by a year, which resembled Endo in conformation but appeared to be more sociable in character, had to content himself with a little family in the human sense – he owned only one mare with her foal and her yearling colt. Because these Fjord ponies had to live in an extremely raw climate, high in the Bayerisher Wald, their condition in the mating season in late spring was rather poor, due to the very long winter in this district. Thus Endo, additionally weakened by covering his numerous mares, was unable to hold his harem together and his brother, Findo, was able to seduce three of his mares together with their complete entourage. After this the two family groups were almost equal, because the stronger Endo unit now consisted of nineteen ponies, whilst Findo led fifteen animals. When summer arrived, the position of ownership of both stallions was absolutely clear and both respected the other's belongings.

At this point, acording to general opinion, one would expect the two stallions to divide their territory and to avoid each other diligently. This, however, was not the case. Although the two families generally grazed separately and owned different sleeping quarters, during the day they met frequently and shared a dozing site and a dusty rolling place, which they visited and used at the same time. During the hot midday hours, they combined to form a single herd and dozed together, so that one could no longer distinguish any form of grouping. I would almost say that they cultivated neighbourliness and good relationships, not only between the mares or the foals, but especially between the two stallions, which spent a considerable part of the midday rest in intensive mutual grooming. This extraordinarily close contact between the family groups caused no enmity between the two sires, which contradicts all the customary ideas about keeping stallions. The close relationship between the two stallions might have something to do with the fact that they grew up together, or perhaps the well-known good temper of the Fjord pony is the reason. Although their behaviour was absolutely peaceful and in accordance with that of the Serengeti zebras, it is difficult to explain. In spite of mutual tolerance, there was a recognizable collective order of precedence between the two families and, in any case of doubt, the lead was taken by the older stallion Endo with his entourage.

The information frequently given in books that yearling entires are driven out of the herd and have to found their own families is based almost without exception on observations made during the mating season, as this appears to attract most of the observers. On the one hand this is understandable, because the herds are subject to a great deal of activity in spring, on the other hand the love-life of the horse is given too much importance in the manner typical of human egocentricity, as we shall see presently. If a mare comes on heat, the family head drives off all other colts to a distance of about fifty yards, with the exception of male suckling foals. At all other times, however, there is a friendly relationship between the father and his sons, which, of their own free choice, will temporarily seek the company of colts of similar age with which they can form a group of their own. A bachelor club of this kind in the Fjord pony herd consisted of a four-year-old small, but fully grown, entire, which could not found his own family but was allowed in the immediate vicinity of the mares by their stallions, and two strong, two-year-old animals in different stages of development, which clearly belonged to the Endo family, to which they remained loosely connected by constant visits. That the latter two returned voluntarily to the Endo family for much of the day was obvious, when, during the midday pause on the dozing site, they mutually groomed Endo. In spite of their fellowship, there was an order of precedence even in the bachelor club, in which the somewhat retarded four-year-old entire had been usurped by the energetic young Asko, which took the lead.

The alliance of a single fully-grown entire and colts of different age groups in an independent group, which often stayed together for a long time, strangely reflects a province of human social behaviour, since in most human societies of different kinds one finds quite similarly constructed bachelor clubs for young men, boys' brigades or whatever these associations call themselves. These alliances of male individuals are due both in Equidae and in some primitive peoples (and until recently in our own society) to the fact that most of these young men, although adult, are not in a position to keep a family. Besides full physical development, a mature personality is essential, both in strength and nerves, for Equidae to engage in seducing a young mare and in protecting her for a lengthy period, so that the 'marriage' is successful. Wild Equidae stallions do not find themselves capable of this at three years old, an age that one nowadays considers old enough to make a riding horse. Instead, as Klingel found out with his Grant zebras, they stay with their own clan until they are five or six years old.

I was able to watch an unsuccessful attempt at founding a family. Asko obviously very much liked a three-year-old mare that had just produced her first foal. Apparently the mare liked him too, since she kept him at a distance only for the first few hours after the birth of her foal and very soon allowed him near herself and her foal. On the same day, two strange horses were rebuffed, which caused great excitement. Asko took his chance, when everyone was engaged with the newcomers, to abduct his sweetheart and her child in secrecy to the furthermost borders of the range. There, high on the mountain, the three passed several hours in the afternoon close together, until it suddenly occurred to Endo that one of his wives and her newborn foal were missing. Unhappily, although the runaways were several hundred yards distant, there was no cover of any sort, so that the deprived Endo saw her at once, darted at full gallop up the slope, and drove the mare and her foal back. The two-year-old Asko made no attempt to do anything about it, for the psychological superiority of the older stallion was such that a slight threat was sufficient to obviate the possibility of a fight. From the point of view of knowledge of behaviour, the frequent use of stallions which are rarely three years old in our breeding programmes is sometimes problematic because of their mental immaturity. Young entires with nervous instability need very careful handling during their first season. I do not think it is out of place in this connection to speak of the possible beginnings of genuine sexual complexes, similar to those arising in people, which can result in a life-long aversion or a lazy mating, premature ejaculation and other behavioural disturbances. Some young stallions give the impression of being inhibited in their behaviour towards older mares, as if their social position were too high for the very young stallion. Our domestic stallions have no opportunity of any kind to practice social behaviour; at most they learn, when out at grass with other colts, to fight for higher or lower rank. In spite of being able to relax in a type of bachelor club, they miss the maturation which wild horses enjoy, namely the constant loosely-knit relationship in a complete family society with a ruling stallion, youngsters of all ages and, above all, mares of different ages. I can prove that, although the young entire in freedom does not have the chance of close sexual contact, he follows the mating behaviour of stallion and mare from the start of her coming into season until the actual mating.

Sexual behaviour is, of course, simply a part of the whole repertoire of behaviour which occurs in natural equine family life. The domestic stallion is also a stranger to the less impressive activities such as how a father behaves towards his small and half-grown children, which a colt learns first from his own experience and later by observation. Though it may sound too human, I am sure that colt foals, which may only spend the first six months of their lives with their sires at grass, have a significantly better-developed, normal, masculine behaviour than those which grow up in the usual herds of mares without stallions. Accordingly, the same applies to the physical development of the mares, for which the association during their youth with fully-grown stallions is just as favourable.

From what has been said it is clear that most of our horses grow up so unnaturally that it is virtually impossible for them to develop a complete and consistent pattern of social behaviour. It is absolutely certain that some 'vices' – which should preferably be called 'behavioural disturbances' – are frequently due to the lack of a father in childhood. With us, too, there are many similar cases and the shortcomings in human social behaviour that arise from some children being brought up alone or in homes are also, in a figurative sense, to be found in horses which had a similar youth.

*Illustrations 50-58*

Plate 21 *above:* The recognition phase is over; the friend is allowed to smell the new-born foal, which is not yet sure where to find the udder and looks for it between the mare's forelegs.
*below:* Scent plays a large part in contact between mother and baby. Haflinger mare Heidi with her foal of a few hours; the mare has not yet cleaned herself and the afterbirth can be clearly seen.

Plate 22 The most important things for a foal are to find his balance and to exercise his muscles. The author's three foals:
*above left:* Holstein foal Grille finds it difficult to get her mouth to the grass.
*above right:* Peppoli, only a few hours old, begins to canter; Notice how she shows her intention by the way she holds her tail.
*below:* Ten-hour-old Holstein foal Aischa circles her mother at the canter.

(Continued page 87)

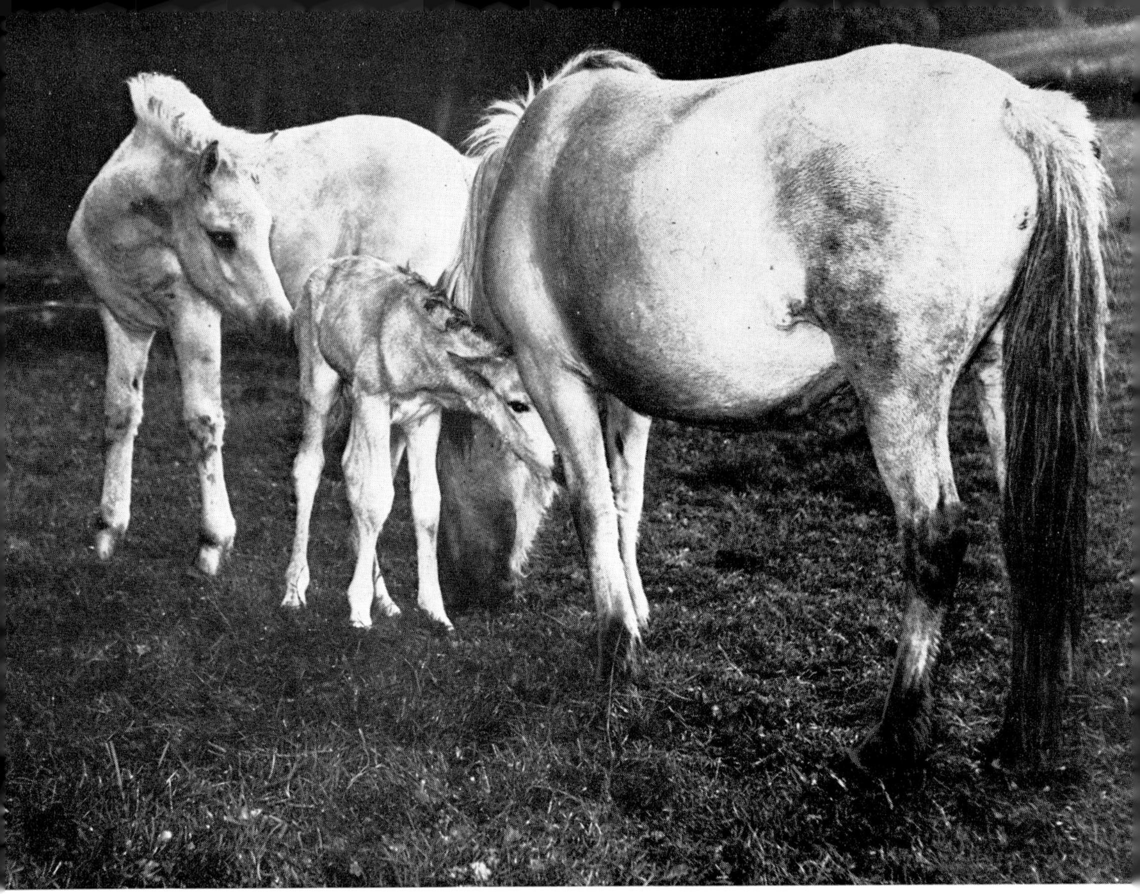

## *Territorial behaviour and greeting ceremony*

Until quite recently it was assumed from the frequent deposits of dung on the borders of their enclosures, paddocks or ranges, that equine stallions were territorial. This opinion must be revised, since most Equidae do not have a firmly delineated preserve, like deer or animals of prey, to defend against rival animals of their own species. In any case, territorial pretensions could not really succeed because the majority of equine species, in spite of their habitat loyalty, are forced by climatic circumstances to expand or contract the extent of their range. This has already been discussed in the chapter on the horse's range, rhythm and pattern of movement. However, a range that any number of families may pass through can scarcely be properly divided. Apart from this, many species of Equidae are at times inclined to form one large herd, which in fact can consist of countless animals – in some zebra species, of several hundred in a fairly small area – which accords with our original concept. Even such a large population would not be led by an omnipotent stallion, but would unite, fairly loosely, a number of single herds, each with its own family hierarchy.

Stallions with followers, which undertake treks together in

Plate 23   *above:* For semi-wild horses recognition is more important than for domesticated for horses. The Spanish pony mare threatens her companion, who must stay further away than is usual with Fjord ponies in the same situation.
*below:* Suckling lasts for nearly a year under natural conditions. Heavily in-foal Fjord mare with a ten-month-old foal.

Plate 24   Games for boys and girls.
*above:* Starting a scuffle – four-month-old Haflinger colts standing side by side.
*below:* Fjord pony fillies indulge in gently nibbling each other's coat.

company, sharing a watering place or grazing a large area, are able to act together under exceptional circumstances, such as taking flight, facing an enemy or causing a rival to flee. One day a fully-grown gelding and a mare in first-class condition were brought into the Fjord pony area. As soon as the strangers appeared, the heads of the families, with their male offspring over two years old, rushed towards them. The gelding, which reacted much like a stallion, tried desperately to keep his rival entires away from his mare by his imposing behaviour and readiness to fight, but without success, and, in spite of his fighting technique and willingness, he was outflanked. In the end the hard-pressed gelding took flight and broke into a sweat out of fear, something I have never seen before in a male horse. The astonishing thing about the whole business, which resulted in the gelding having to graze several hundred yards distant, was the fact that the two family stallions co-operated to drive him away and the two-year-old colts joined in this unfair battle and, although they did not actually attack him, they certainly threatened him. At the end of the drama one might have expected a fairly serious battle between the two stallions over the new mare, but, quite as a matter of course, she was appropriated by the higher-ranking Endo, whilst Findo, for his part, made no attempt to stake a claim and returned to his own family. It is interesting to note that, in spite of all the battle cries and truly menacing turmoil, not a single horse was hurt.

At large gatherings of many non-territorial groups of Equidae, familiar and unfamiliar heads of families automatically and repeatedly inspect each other, during which time a rather extraordinary ritual takes place, which Klingel has described as a mutual greeting of the stallions. On the other hand, I would not subscribe to Klingel's description, as in my opinion there are two quite different behaviour-complex reasons, namely a 'tabooing' of the family society and ritualistic territorial behaviour, which have a single outcome.

If two family stallions approach within a certain distance, which with Fjord ponies is about one hundred yards, they leave their mares behind and go to meet each other at the trot, with their ears pricked, in an attitude calculated to impress. They sniff each other on the nostrils, ears still pricked attentively; then, as a symbolic attitude of aggression, the ears go back, necks are arched and both horses stamp the ground energetically with a fore foot. The whole performance is accompanied by loud squealing, which may increase almost to a roar. This show of strength – the stamping and the bellowing which cannot be compared to the normal neighing of a stallion – has a definitely frightening effect on the person watching, who may well imagine that a battle between the two rivals is about to commence. However, nothing of the kind occurs. After the first demonstration of individual power and splendour, the horses go round in circles side by side, while each stallion sniffs the flanks of his opponent, finishing in the region of the genitals. Extraordinarily enough the head-to-tail position is precisely maintained during this ceremonial performance. It seems to me, for this reason, to be similar to the position when coat scratching occurs and symbolic, therefore, of absolutely friendly intentions. At the end, the highest ranking stallion takes a step to the side in order to deposit some droppings, which the second stallion for his part then marks by staling or dropping dung. This solemn action is always completed by a mutual sniffing of the excrement, whereupon the rivals, giving a slight threat with the ears slightly laid back and the suggestion of a rear, turn away and return to their families. This ceremony is repeated during the day whenever two stallions get too near each other at any spot on their range.

When the ritual has been accomplished, the families may wander off, but sometimes they may continue to graze side by side and form a mixed community in some shady place during the midday dozing period. That a fight does not take place and that eventually a certain period may be spent in company are two very interesting and extremely important facts as far as the whole social life of horses is concerned, because they make the formation of a proper herd feasible. Throughout the ceremony described, which apparently has to be repeated frequently if the effect is not to be lost, the mares are clearly so taboo for the other stallion that sometimes there is no need for an appreciable distance to be left between the herds. This is my experience with the Fjord ponies and Klingel's observations of the zebra species show that a stallion, when keeping to the rules of the game, will not in any way seek to molest or abduct a strange mare. The reduction of the distance between groups or individuals can even be such that mares at the dozing site from different families may even touch each other and the stallions may indulge in mutual grooming of the skin. I think that the sight of two fully-grown herd stallions nibbling each other's coats in the presence of all their followers must be fairly sensational. Logically, mares that know each other do not do this.

Taboo in the family communities is so effective that even sick stallions cannot be displaced at once by a healthy rival. One day Findo was extremely lame behind, which made him totally unable to fight but, in spite of this, Endo made no attempt to carry off any of the mares, even though originally some had belonged to him at the beginning of the year. This behaviour coincides with Klingel's observations of the Grant zebras, whereby he established that, despite the death or elimination of the leader through age, the deserted harem kept together of its own free will, until it was taken over by a bachelor entire or by another family stallion with only a few mares.

Besides this family taboo, which alone makes a large social

community possible, the second purpose is a form of agreement to use a larger territory together. It has already been mentioned that horses which daily roam a wide range in groups following each other generally have no particular area that they mark or defend. Almost all Equidae have this pronounced territorial behaviour, which was probably common to the forebears of today's Equidae, millions of years ago in their tropical forest habitat, and had to be relinquished as useless during the process of evolution. Since habits of behaviour die slowly and are much more permanent than physical characteristics, in the animal world they are ritualized in a few symbolic acts in order not to disturb or prevent the survival and continuance of a species. The ceremony of the stallions upon encounter, with its symbolic fighting, elements of imposing behaviour throughout and mutual marking of droppings, still shows this earlier function of defending the preserve and the family.

The ritual is, of course, not proceeded with if a herd stallion meets a young entire that has no mares of his own. At an early age young entires appear to practice single parts of the ceremony — the marking of droppings is especially pronounced at an early age – and even the fighting and imposing behaviour are often practised without necessarily resulting in a fighting game. It is very obvious that they are not only physically much more energetic than young mares, but give a definite impression of being mentally more alert than their female companions, because they are interested in everything new. For example, they are terribly interested in each new-born member of the herd and they stay in close contact long after the herd stallion has wandered away, which is the opposite of the young mares or fillies, in which the eternal motherly instinct and a corresponding liking for baby foals would sooner be expected, from the human point of view.

## Special forms of social behaviour

One is inclined to transfer the results of new and interesting experiments to other similar animals with much the same habits without examining in detail their analogous conclusions. Up to the present we can confirm with certainty that the pronounced and peaceable social behaviour of the Fjord ponies already mentioned corresponds almost exactly to the behaviour of the steppe zebras and their tendency to develop large herds, which Klingel investigated, and we can suppose that this is also true, in the main, for ponies of similar conformation. Klingel drew far-reaching conclusions concerning all genuine horses, although he relied entirely on written records about pony populations and on his own studies of the same. His current researches on further wild Equidae have, in any case, already produced some results which are quite consistent. They are largely connected with the conformation characteristics of these Equidae and demonstrate distinct parallels with the domestic breeds of horses which I had investigated, which generally did not belong to the pony types. The result, for example concerning the Grévy zebras and wild ass stallions, shows that they are still partly territorial and, by establishing marked dunging places, own defined preserves. The latter conforms significantly with some of the characteristics and primitive anatomical peculiarities of these Equidae.

Grévy zebras, which only defend their preserve when there are mares on heat in the vicinity, already show a tendency to the purely symbolic territorial behaviour of the Grant zebras or our Fjord ponies. Their social structure and hierarchy also appear not to be so significant, and social unity is obviously less pronounced, even if it sometimes consists of several hundred animals gathering for short periods.

A fine-boned, wiry stallion of a mouse-dun colour was introduced into our herd of Fjord ponies. He looked much like a

Tarpan, reminding one of an Arab, because the owner of this breed was of the opinion that Tarpan-style animals ought to be the most suitable for keeping on a range in the cold Bayerisher Wald. The experiment was in the long run completely impractical, not because the stallion had no masculine qualities or could not have stood the winter, but because his whole behaviour was so different to that of the other horses and, above all, to that of the other stallions so that his stay in the herd posed too many problems. This Tarpan-type newcomer showed, amongst other things, fairly pronounced territorial behaviour. He was constantly in action defending an imaginary boundary going straight through the range, which had nothing to do with the lie of the land, and thus completely prevented the cycle of roaming of the various groups through their territory. Frequently, too, he infringed the taboo of the stallion ceremony and let it degenerate into a real fight. In many ways this stallion even conducted himself differently towards his own family. This fact could be recognized by the mannerisms of an overlord, so often described in literature, since he was continually circling his mares, keeping them unusually close together, and did not allow his harem to mingle with the mares of strange stallions. This may well be a matter of individuality, since peculiar differences appear in other breeds of horses as well. Some Eastern horses, including Egyptian Arabs, possess certain original characteristics of conformation and consequently some archaic traits. It seems to me that this strain or type of horse has retained primitive territorial behaviour. I do not rely simply on studies of Fjord ponies but on numerous additional observations of different, well-bred, warm-blooded horses and especially on observations of the completely feral ponies of northern Spain. These ponies have a far larger reserve, free of boundary fences, to range over; some also possess fairly obvious oriental characteristics and the herd stallions show the same energetic assurance of ownership as did the Tarpan-type Fjord stallion. Although there was never a proper quarrel, not even in the spring when mares on heat were constantly present, the stallions kept an obvious distance between their families which was much more severe than that of the Fjord ponies or the zebra communities described by Klingel. As a last example, I would like to mention the forest Tarpans in the Bialowiecz Forest which also apparently have a pronounced territorial disposition. These horses in the Polish reserves were once to be found throughout south-east and eastern Europe.

There are analogous instances in the behaviour of wild asses and our domestic horses. As far as it is known at present, wild ass stallions are to some extent territorial, although they do not form large herds and spend most of the year alone. I discovered similar behaviour amongst some warm-blooded horses. It is well known that raw-boned, ram-faced horses* generally have difficult temperaments: they are nervous and their characters are easily spoilt. This view is, I think, generally the result of cowardice on the part of people, because they do not understand this type of horse; even so, there is some truth in the matter, since many horse-lovers cannot cope with them. The 'ideal' type in Germany, probably for historical reasons, is the sturdy well-made type, represented by many European pony breeds and, indeed, by the warm-blooded and heavy breeds. Their pleasant, extrovert temperament and patience make the contact between man and horse so much easier. In the United Kingdom the 'ideal' type is probably the Irish hunter – which can be used for so many purposes – and the Arab or Anglo-Arab. The latter breed is popular in the U.S.A., though there, of course, the people of the various states have their own 'ideal' breeds.

The reason for the difficulties which are especially apparent in these overgrown, rather aggressive, convex-faced horses lies

* Translator's note: Horses with convex profiles.

chiefly in their less tolerant social behaviour, the cause of which may be found in their original environment. Like wild asses and probably some Asiatic *hemionus* varieties,* they needed a distance essentially larger between one animal and another than many other Equidae, because of the sparse grazing in their original habitat. For this reason, even though they adapted in many ways to new conditions, this larger distance between individuals became part of their behaviour pattern and obviously has not changed in the relatively short period of domesticity. The looser connection of the herd members of this particular type of horse used to be a requisite characteristic of the remount in the days of cavalry, because at a moment's notice it might be required to leave a troop and go off alone.

This type, which has an inherited need for personal space, can be fairly frequently seen in the corresponding warm-blooded horses. Such animals are impossible to handle once this space is curtailed or if one forces them to live in a cramped area. Usually they graze alone or only with their foals and they do not like either herd life or the friendliness of mutual skin grooming. In the stable they often kick the sides of their boxes, especially when they have neighbours. The continuous effort of these horses to remove any other horses that are too close creates not only an extremely annoying noise for people to contend with, besides damaging the box, but it can cause injury to the animal itself. Thus the box of E. V. Meindorff's Andalusian stallion, Jaguar, had to be padded, as this stallion, although ideal for High School because of his excellent carriage and impressive appearance, was forever kicking the boards.

Even the specifically independent behaviour of these stallions reminds one of the wild ass, because for most of the year they lead a solitary life and, apart from the mating season, keep a fairly loose and distant contact with the mares. Although one

* Translator's note: Onager, kiang, kulan, etc.

often sees these horses in jumping competitions, Olympic trials, etc., they are generally too valuable for more than one stallion to be kept on a free range. For this reason I unfortunately could not investigate the behaviour of adult male animals in a group *in situ* and thus these remarks are mostly the result of theoretical inferences. However, I did have the chance to study the behaviour of a single stallion of this type, which later became one of the top show jumpers, throughout his period of development. As a yearling this stallion was turned out on a large paddock beside the one in which my mares grazed and, even when fully grown, he only occasionally grazed near the fence, but generally he kept his distance of several hundred yards from the mares, and showed no herd instinct, and no desire to get to the mares. If a mare was in season, he did play the full repertoire of stallion behaviour and had to be removed from the paddock, otherwise he would have jumped the fence. Similar conduct has been observed in some onagers, which possess many characteristics of conformation typical of the markedly running animal. When wild or semi-wild Equidae move to another area, they wander, as described, in single file in family units behind the lead mare, whilst the stallion observes no particular order and is generally found at the rear of his family, although he may frequently be keeping watch to the side for enemies or to encourage the herd to change direction. In the order of hierarchy the lead mare comes immediately after him; she is the highest ranking female animal and, if there is no danger, she determines the direction to be taken. Amongst herds of horses in which the stallion runs with them for only a short time or which consist only of mares, the position of the lead mare is much more dominant than in the natural herds of Equidae. The well-known hippologist, Ebhardt, observed that, amongst the semi-wild Icelandic ponies of the heavier 'cart-horse' type, the influence of the lead mare was so pronounced that, when the stallion was turned out in the spring, such a violent battle for precedence took place between the two that Ebhardt thought it could actually be called a matriarchal society – at least amongst some Icelandic ponies. I would not like to go so far, although my own observations of the commoner warm-blooded horses seem partly to confirm this. Possibly Equidae have patriarchal and matriarchal social habits running side by side, and these overlap or supplement each other just as in human societies.

If there is no stallion running with a herd, an especially high-ranking and aggressive mare may assume the duties of the stallion, and another mare will become the lead mare. In my own herd the extremely active five-year-old Aischa occupied this position: she generally grazed alone and when the herd returned to the stables at dusk, she always acted as the guard and waited until all the members of the herd had gone through the gateway. She occupied third place out of eleven fully-grown animals in the order of precedence, so that she would never normally have been the last through the gateway, even though she had to let others go first. The clearly masculine characteristics of this mare, which had, in fact, produced two foals, were also seen when she separated an in-season companion of lower rank from the rest of the herd and tried to jump her, which, contrary to cows, rarely happens amongst horses.

The habits of social behaviour described below, which unfortunately I did not witness myself, are taken from Klingel's experiments in order to complete the picture. One of the most astonishing results must be the stability of the social groups of the Grant zebra under investigation. Apparently the adult animals generally remain together all their lives and only the young animals leave the family circle. The unity of members of the family is so close that a new animal trying to enter the community will not be accepted for days or even weeks and its integration is finally only brought about by the intervention of the stallion. Thus, in the absence of the head of the family, a strange stallion could not abduct one or more mares, because they would not want to follow him and might, indeed, repel him. If the stallion falls victim to a carnivore, the mares continue to stay together, until the whole group is adopted by another stallion. The normal replacement of a very old or weak family stallion happens gradually, as a rule, over several days and generally without any fighting; this cast-off animal may join a bachelor group, which frequently consists of his grown-up sons, or he may continue his existence independently.

Fillies leave the family at the age of eighteen months approximately, when they come in-season for the first time. This is by no means a voluntary action, in that the family neither lets them go nor are they looking for freedom; they are taken away from their sires by strange stallions. They show their first heat, which lasts an unusually long time (as with domestic horses), by a characteristic carriage. This acts as an immediate signal to all bachelors and family heads which possess only a few mares, and generally encourages a number of suitors. Scent appears to play only a small part in this mainly visual signal. The stallion certainly defends his daughter with all his might, but he generally succumbs during the week-long heat to the overpowering number of eager suitors – once eighteen were counted! After the filly has been abducted, the battle between the kidnappers continues, until the mare finally becomes part of either a newly-formed or an existing family. Sometimes the father succeeds in warding off the suitors and the youngster remains in her own family group. In most cases, however, the abduction of the filly is successful and the rate of inbreeding under natural conditions is kept fairly low, even though inbreeding or incest in the form of father-daughter mating could in theory happen relatively frequently. An instinctive barrier to prevent such close inbreeding does not appear to exist, as some owners of horses suppose.

The social behaviour of Fjord ponies, within the sphere of my observations, almost exactly matched that of the Grant zebras. I believe, therefore, that conclusions can be drawn from the analogy with a fair amount of accuracy both in respect of founding new families and casting off the family stallion. Further research will show how far similar discoveries about zebras and ponies can be transferred to other types or strains of horses.

# 4

## SEXUAL BEHAVIOUR

There has never been an aspect of the behavioural habits of the horse so much written about as its sexual life. This is probably due to the fact that it is easily observed at the birth of a foal and when a mare is on heat, whether in reserves, zoological gardens and breeding establishments, in freedom or in studs. Besides the mother-child relationship, the mating act, even for the layman, is so clearly recognizable that it offers the ethologist a theme to study. In addition, I think that many people possess a certain voyeurism, whether they admit it or not. Schiller said of humanity that 'hunger and love make the world go round', but the consideration of the sexual behaviour of Equidae has received too great an emphasis. As far as hunger is concerned, what Schiller said is probably true, because eating is actually their most important occupation, but, concerning love or sexuality, the statement is valid only within a very narrow compass. With human beings eroticism is an aspect of our behaviour during the greater part of our lives, an aspect in which reproduction plays an incidental rôle. In other animals with an intellect, like the apes and especially dolphins, its social function is evident. This is not the case with horses: sexual behaviour is usually exercised to serve reproduction and plays a very small part in the sphere of communication.

However, there is quite certainly a kind of love between

Equidae, which, though they may not be aware of it, is without doubt the cause of many actions, such as the way that female members of a family choose to live close together and the friendly relationship between stallions and their sons or between mares and their foals. Even the equivalent of human affection for a partner is present, otherwise how can one explain why some stallions with a wide choice have a favourite mare, which they prefer in such a way that one cannot speak of a simple hormonal or vegetative reaction? These preferences cannot be explained simply in terms of hormones or custom, just as such feelings in man cannot be explained in this way. How similar the higher developed animals are to humans especially in this elementary sphere of life, to which not least sexual behaviour belongs, is seen from the vernacular statement, 'It's only human'; 'human' in this sense qualifies actions which have a purely physical motive and which in no way make us superior to the animals.

## NATURAL MATING

There is a fundamental difference between the sexual behaviour of a stallion and that of a mare. All Equidae stallions are always ready to mate; they have no special mating season. On the other hand some mares come on heat only in spring and early summer; in southern regions the heat period is occasionally spread over the whole year at three-week intervals according to surroundings and species. Their actual willingness to mate is exclusively at the time of the release of the follicle. The heat cycle varies with individuals and sometimes lasts several days. Domestic horses usually come in-season at the age of one and a half years depending on the breed and the state of nourishment; warmer weather and longer daylight hours are also important factors. After an interval of seventeen to twenty-one days, the next heat cycle should follow.

In intense sunshine the heat is especially strong and the level of fertility is at its highest; these lessen towards late summer and in the winter months they are generally completely absent. These favourable periods for conception ensure successful reproduction, because, after the eleven-month term of pregnancy of genuine horses (as opposed to asses, zebras, etc., which have a different pregnancy period), the foal will be born in the warm sunny months, when grass is at its best. Even among stabled horses, the fertility rate of mares covered during the natural mating months is considerably higher than in those covered in late winter or very early spring. This problem presents certain difficulties for the breeding of racehorses, since foals have to be born during the early months of the year. Their age is counted as from 1st January and, regardless of whether the foal is born in late spring or late autumn, it becomes a yearling the following January. It is only possible to bring two-year-olds to the starting post if the youngster is really two years of age and even a few months older. Indeed, these Thoroughbred or Trotter foals are not just produced to order, since some clinically absolutely sound mares, which are covered too early, continue to come in-season, until they get the necessary amount of green fodder and above all probably the requisite ultra-violet

*Illustrations 59-66*

Plate 25   *above and below:* For want of a companion of the same age, other animals are accepted as playmates – Leonberger bitch Orlette with Intrigant, a Trotter colt.

Plate 26   Fighting game between two colts.
         *above:* The Arab stallion Neshar tries to bite the Haflinger stallion Freiherr, alias Motzl (75% Haflinger, 25% Arab blood).
         *below:* Start of a rearing battle.
(Continued page 103)

rays. If they lack the benefit of these rays, I call this the 'U.V.R. brake'.

The long period of pregnancy results in the obvious necessity of a further mating shortly after giving birth, if the next foal is to arrive in the spring. Generally mares come in-season again seven to nine days after foaling, although after a difficult birth, or at an older age, it may be as long as twelve days or more. It may appear to the outsider that nature is rather cruel in causing a fresh pregnancy so soon, but it is essential for the continuance of the species and is practised, too, when covering is done by hand, that is to say when the breeder controls and helps with the mating. Thus foaling would normally take place twenty days earlier every year but, so that the birth terminus does not occur during the cold months, the U.V.R. brake prevents the mare from conceiving. This preventive factor can be eliminated, however, by feeding the mare vitamins and some hormones for a given period. Amongst the Grant zebras, which live near the equator with the same amount of sunshine all the year round, foals are born during the whole year. On the other hand, the Fjord ponies in the Bayerischer Wald usually have their foals in late spring and early summer, although both stallions are always running with their mares and could animate them to come in-season at any time; the extremely tough climate seems to be more effective than the sexual stimulus of the male animal.

Plate 27 Additional phases of the fighting game between two-year-old stallions.
*above*: Neshar tries to avoid the attack.
*below*: Attack and defence.

Plate 28 *above*: Symbolic fighting game between man and horse. C. H. Döhmken plays with his pure-bred Arab stallion Mahomed.
*below*: Quadruped Test. Hungarian horses in Leutstetten flee from a man on all fours.

The stallion is aware of the mare's approaching heat before she is actually ready to mate. All male Equidae are interested in the droppings of their companions throughout the year and devote much attention to the dung, etc., of their own mares to ascertain if the mare is approaching her heat period. During the few days before, the stallion will pay a great deal of attention to this particular mare: he always grazes nearby and is continually nibbling and sharing care of the skin, and quite early unmistakably offers her his attentions. This is all very carefully planned. The suitor approaches in a more or less pronounced 'showing-off' manner, an attitude calculated to impress, with arched neck and a collected high trot. The more familiar he is with his loved one – perhaps a long-term member of his family – the less he tries to impress her. The less he knows her and the stronger she appears to him, the more he shows off and tries to emphasize his best points. After the stallion has repeatedly sniffed at her shoulders and flanks, he turns his attention to the more important sexual organs and will *flehm* with abandon. We shall return to the *flehm* gesture. During the smelling, he exposes himself and gets ready to cover her.

At the commencement of the heat cycle all efforts to get closer are generally warded off with vigour. The mare will threaten her admirer with ears laid back and by kicking with one of her hind feet. If this obvious repulse is ineffective, she will then kick out with both hind feet, uttering angry squeaks and ejecting urine, so that for a time he retires, although staying near her and keeping her under observation. In books this defensive behaviour is described in a detailed and generally dramatic fashion, in such a way as to suggest that a mare, which is not yet fully in-season, attacks the stallion kicking him in the jaws. However, I have often witnessed a pestered mare giving vent to her feelings with some force, but there was never a question of her kicking him badly about the head, because even the most inexperienced stallion would have learnt how to avoid this during his youthful fighting games. At the most, I imagine that a completely 'innocent' young entire, which had been brought up alone, might possibly once be caught on the head, because his reaction in avoiding such a kick was too slow. Although there were many youngsters and numerous adult entires of various breeds from Shetland pony to Lipizzaner running with herds in my care, I have never come across any that were ever kicked on the jaws or head. If a mare does kick then the already alert stallion would quickly lift his head and avoid the blow by moving sideways; if he were too near, he would rear up, take the kick on his strong chest muscles and would come to no harm. Neither his belly nor organs are actually in danger, since a free semi-wild stallion and experienced family head would never try to take sudden advantage of his wives as a stallion on the lunge with a groom in attendance might do. Sometimes a fiery young colt with no experience whatever, suddenly finding himself running with a herd and full of pent-up energy, may get hurt under certain circumstances. After the first shock of a mare's defensive action, even if he has not been hurt, he soon learns to be more careful and discreet in his tactics. An experienced stallion watches the reaction of his mare very attentively and, even when she kicks, he can judge just how far the kick will carry, so that he need move away as little as possible.

A normal heat lasts only two to three days, during which time a healthy mare changes her attitude. Her behaviour usually clearly shows whether she is coming into or is fully in season, or not. When she is ready to accept the stallion, she will stand with a sway back, hind quarters slightly sunk and legs spread, the tail held out and head forward, thus she can attract a stallion even if he is some way off. The ejection of urine and mucus, which signals the change of metabolism, has an erotic effect on every stallion. In a very short time, after a few passes, the stallion will serve the mare. The act of mating may occur several times during the day, sometimes at very short intervals, and this can tire the stallion.

If several mares come in-season at the same time, which is very likely with a small semi-wild herd, one or other may not be served so often and the stallion becomes less active. In addition he frequently has a favourite mare, which he especially wants to court, and sometimes she can simulate her readiness to mate before she is actually on heat. Such services are infertile. Breeders of Icelandic ponies, who are probably the most experienced in western Europe at keeping semi-wild herds with natural mating facilities, give the stallion a maximum of twelve mares, sending him to roam with them in the mountains unsupervised. It has been proved that a larger number of mares will not be one hundred per cent fertile, unless some assistance is provided from outside. Once again it is confirmed that the small family group is the normal natural size for a successful stallion, as we have seen with the Fjord ponies and as has been observed with zebras.

I have already pointed out that probably a part of sexual behaviour is acquired, since the stallion's courting of the mare, his gentle coaxing by nibbling and the actual act of a mating are carried out in the midst of the family group, in very close contact with the other animals, which follow these activities attentively. One can often see a week-old foal getting in the way of its parents and under their feet. Nervous horse-breeders are often afraid that something could happen to the foal but, at least, a stallion that is always running with his mares is in no way inhumane, on the contrary, he is usually a kindly father to his children and, if he wants to occupy himself intensively with the mare, he will generally push the little one aside carefully with his nose. Whilst the stallion is actually covering the mare, the foal usually stands in front of his mother. He finds the whole event rather strange, possibly because he no longer recognizes either of his parents as individuals in this position. The expression on his face is one of timid submission.

If it is a complete family group and half-grown animals are also grazing together then they, too, are very interested in this event. Fillies can get so close that they smell their dam's noses, but the colts are kept at a respectful distance of several yards by the threatening attitude of the stallion. If several stallions are running with their families on the same range, one might perhaps expect that fearful fights would ensue as soon as a mare comes in-season, but, up to now, I have never seen this in small genuine communities. The male animals are chiefly occupied in keeping their mares away as far as possible from rivals. In general the property of a neighbouring stallion is respected, and no other stallion is inclined to try to seduce one of his mares when he is in the process of covering another.

Most scientific publications about the mating behaviour of Equidae are based on the study of primitive horses or ponies and their social structure, which is almost as unnatural in their reserve as that of herds of mares kept at studs (as in Germany, Hungary, Poland, etc.). Some of these primitive ponies which have been investigated, like the Dülmen in the Mehrfelder Bruch, are under matriarchal domination and generally a single well-fed and well-prepared stallion is turned out with them in the spring. His remarkable behaviour is seen by his preoccupation with trying to unite the separate family groups in one large herd. With a threatening expression, his neck and head almost on the ground, ears laid back, tail in the air and an almost angry snaking action, he constantly circles the herd to prevent single animals, or even groups of mares, from breaking out. Almost all male animals of various origins and temperaments behave like this, if they are placed in a community that is too large for them to manage alone, where, instead of real rivals, they have to contend with a hierarchy battle within the group of mares; the extent and vehemence of their aggression varies with the individual, depending upon the stallion's potential energy and despotism.

This astonishing aggressiveness, so often described, is supposed by many writers to be one of the preliminaries to mating. I do not believe that this is so at all, since it is almost completely absent in large herds consisting of family groups that have grown up harmoniously together, or at most it is only hinted at when a mare is about to come in-season and is some distance from her own group. Some stallions may assemble their harems with varying degrees of energy much as when communities which are strangers to one another suddenly meet or when a very large group of bachelors is in the vicinity. If a herd sire wants to change the direction of his wandering column of mares and foals, he will canter alongside making threatening gestures, so that he forces them to turn aside. Even a stallion which has been used for riding and then turned out with his mares in a paddock canters around in this way to make sure that everything in is order, but he stops as soon as he is satisfied that he has nothing to worry about.

Individual members of the Equidae species appear to have accepted this prelude in varying degrees as part of their sexual behaviour. It is particularly observed of the Grévy zebras and asses that their mares are only willing to mate after they have been chased for some time. This could be connected, especially in the stallions of these two species, with their individual territorial instinct, but we must wait and see how far the present investigations confirm the descriptions of earlier scientists. The diverse origins of our many domestic breeds of horses also produce a slight divergence of sexual behaviour and may be responsible for the large individual variations in this sphere. Driving and chasing do not, in my opinion, form part of the peaceful mating of genuine horses and are resorted to only in exceptional circumstances. However, the circumstances are exceptional when a stallion is let loose in a herd without one and this places him in a situation which is too demanding, since he is not like a wild zebra colt seducing a filly in-season or conveniently taking over a complete family of two or three mares; all at once he has to rule numerous closely-knit families. Perhaps his exaggerated conduct is even the result of early behaviour when carrying off a filly, which he then transfers to this occasion, for which Nature has not equipped him. He tries to obey his instincts as best he can in this difficult and unnatural situation. It sometimes happens during the spring in these domestic free-range herds that several mares adopt the unmistakable heat position, then approach and pester a stallion. Generally, in these cases, he does not divide his favours equally and clearly prefers one of the more attractive mares, which can cause much importunate conduct by the other in-season rivals. Mutual threats and battles for precedence are likely to result from what one might call sexual jealousy, so that constant commotion and squabbling, accompanied by squeals and kicks, are a delight to the watching behavioural researcher. In the meantime, the stallion does not behave in a gentlemanly manner and often violently threatens the mares that do not appeal to him, even driving them off, as long as he is occupied with his favourite. Should he roll near her, then his lady-love stands there, sometimes with a really silly, not to say childish, expression, like a foal that sees its mother rolling for the first time; even in horses, love appears to affect their mental faculties!

## COVERING BY HAND

Using the stallion on the lunge and covering by hand have already been mentioned, and this is the usual method of mating a mare at stud. The pre-mating ritual is almost ruled out and the whole process of sexual behaviour is reduced as far as possible to the simple act of mating. Since there is no chance for the stallion's natural playful approach to the mare before she actually comes in-season, after which both are ready to mate, one has to discover if the mare is in fact ready. Nowadays large Thoroughbred and Trotter studs carry out a manual veterinary examination internally by which the mare's condition can be diagnosed, so that the most suitable time for the actual mating can be decided, but generally studs still use a teaser. The mare is led to a padded trying board in the covering yard; this is necessary to protect the stallion from being kicked by a mare which is not quite ready or which may even be neurotic. A teaser is usually used – a horse which is very quiet, with a nice temperament and easy to manage – and he is encouraged to smell the mare's shoulders, flanks and hind quarters, so that her reactions can be watched. If she wears the 'mating expression' and stands with her legs apart, lifting her tail, the clitoris will be twitched and, if she shows no inclination to kick, she is probably ready to mate and the teaser is taken back to his box, and replaced by the chosen stallion. The stallion should be allowed to spend a little time courting the mare before he actually covers her. It is advisable to have a practised stud-hand or stallion man in charge of the stallion, with two helpers to hold the mare. It is inadvisable for amateurs to take charge of a stallion unless they are prepared to learn stud-work and the care of a mare and stallion before, during and after mating has taken place.

It is preferable to take time to give a mare confidence, rather than to resort to hobbles or the twitch. Maiden mares, which have never seen a stallion before, should be given plenty of time, and be treated very quietly and with patience. It is most important for the future success of the young stallion in carrying out his service with the assistance of stud-hands that they should treat him properly and sympathetically; he is certainly born with the ability to serve a mare, but at the beginning he obviously has no experience. The early impressions which he gains from his first service are a determining fact regarding his freedom from complexes later. It is scarcely necessary for me to add that a mare should be covered several times during heat. If a particular stallion is very much in demand or if it is usual to let him serve a great number – in Germany it is not uncommon for one stallion to serve between eighty and a hundred, and sometimes more, mares – then it is necessary to maintain fertility by intensive feeding with high protein and vitamin concentrates.

## 5

# MOTHER-CHILD BEHAVIOUR

### PERIOD OF PREGNANCY

If the mare is in-foal, she will have an eleven-month pregnancy. No precise period of pregnancy can be given, as it varies between 320 and 355 days, during which time 95% of all mares will foal down, though on average there are only 15% of births between 335 and 337 days, the middle of the period. Apart from this wide variation, which may be reduced or exceeded, there is the added complication of the normal repeated services – the last may be a week after the first service. Probably both the quality of fodder and breed characteristics may have something to do with the length of a pregnancy, in so much as cold-blooded mares which mature early have a shorter pregnancy than many eastern European native breeds, while warm-blooded and Thoroughbred mares come somewhere in between.

Besides the doubt due to breed or, more precisely, the type differences in the duration of pregnancy, external influences play a very considerable part. After careful examination, some characteristics attributed to a breed turn out to be due to a fault in stable management and the pregnancy may become shorter if one keeps the mare under better conditions, including not only a sufficient quantity but also the best quality food, with the necessary content of protein, minerals and vitamins. Added to this and of great importance are exercise in the fresh air, sufficient light, and good grazing during the summer and autumn months on high-quality pasture grown for horses as opposed to cattle. Both horses and cattle do well, however, if grazed together or if horses follow cattle. It is frequently noted as a matter of interest that a pregnancy is shorter if the mare is covered between June and October, and by contrast the pregnancy is longer if the mare is covered between November and May, during the months when there is little sunshine. I, personally, think that cause and effect have been misunderstood, and that it has nothing to do with the date of covering in the summer months but is due rather to the effect already mentioned of ultra-violet rays during late pregnancy, when the mare is out to grass.

In my own stock, about which dates have been carefully kept and registered for many years, the length of pregnancy lies within the lower register; this is due to the proper amount of food throughout the year and the correct exercise. In spite of this, there remains a wide variation from animal to animal, which can be classified into two periodic groups: the mares which transmit size and bone, and produce big foals by every stallion, need on average approximately twelve days longer than the mares that produce small offspring even by large stallions. This fact appears to agree with the opinion held in racing circles that racehorses with big frames mature late and, therefore, must be carried longer than the small strains which mature early. Some fifty years ago the well-known German hippologist, Von Oettingen, made a significant remark, that Thoroughbred foals born after a fairly short pregnancy turned out to be good racehorses, whilst those that were carried longer never developed into high-class horses. Besides the possible connection with the present type of Thoroughbred mare, which can vary considerably, I think the principal factor is the vitality and maturity of a mare, which are transferred to the unborn foal in its mother's womb, and therefore cause the foal's urge to be born. The length of pregnancy depends also upon the mare's age and is shorter with the first-born and in young healthy mares.

Colt foals are usually carried a day or two longer than fillies and thus, providing one has exact knowledge of the usual length of pregnancy of the mare, if she has exceeded her average time, a colt foal can be expected. Ebhardt made an interesting discovery on how to detect the sex of an unborn foal in his Icelandic ponies, which I confirmed once with Haflinger ponies on free range and which was verified by witnesses. Ebhardt writes: 'In the spring when the older mares are heavily in-foal, the stallion is removed from the herd and is given a three-year-old filly to cover – outside the reserve, naturally. When the freshly-covered filly returns to the reserve, all the other mares become very inquisitive and get wind or scent of the filly. Their behaviour varies considerably. Some of the older mares begin to chase the filly, biting at her dock, and give out muffled whinnies and take on masculine characteristics. The rest of the older mares approach within five or six paces of the filly, getting wind of her and remaining very shy. A year or two ago I gave the Rector of the Veterinary College in Hanover an explanation of the two different attitudes adopted by the older mares, emphasizing that the hormone count of the older mares was influenced by the sex of the foals: the mares which behaved in a masculine way carried colt foals, whilst the others carried filly foals. Students of the Veterinary College, Hanover, studied my predictions for three consecutive years and found them to be confirmed, without exception, as soon as the foals were born.'

In-foal mares, especially young ones, become considerably

quieter, even before they begin to increase in size, and they lose their sometimes childish or 'teenage' habits – they are no longer quite so skittish and, compared to companions of the same age which have not been covered, their manner is much more adult. If one knows one's animals well then their condition is confirmed by this changed behaviour, even without a hormone or manual examination. In-foal mares do not want to run and play so much, and during the second half of their pregnancy they rarely exercise themselves and then only in a rather forced gait. Although individual variations do happen, the exceptions prove the rule.

Hunger grows in proportion to the growing embryo and, when it is possible, this hunger should be satisfied. The embryo grows most during the last three months of pregnancy. This is noticeable in the mare's body; her drinking habits change and cold water is taken with hesitation or at short intervals, as it produces a kind of cramp in the mare's stomach and very often causes the foal to move, as can be seen by watching her flank. Approximately two months before it is due to be born, the foal appears to turn and to change its position, and at this time the mare sometimes appears to be a little unwell, eats badly and may show signs of mucus tinged with blood. However, it is still not possible to give the exact date of the birth.

As the mare's time draws closer, the ligaments of the pelvis begin to slacken and her udder increases in size. Within forty-eight hours of the birth, one will notice a small amount of thick, transparent, serum-like fluid from each teat canal – this is know as 'waxing'. Many mares begin to sweat on the neck and shoulders when parturition is imminent, and they become very uneasy and restless, turning round and round, and even kicking at their bellies. A peculiarity of all Equidae mares is their ability, if necessary, to delay parturition for several hours and this has several advantages for wild mares. In domestication, this strange ability has been retained and mares will wait until all is quiet, usually at night, before giving birth to the foal. Statistics show that foals are usually born between 8 pm and 7 am. Wild Equidae have to be constantly on the watch during the daytime, so it is probably at night that they feel safest and least likely to be disturbed in their habitual sleeping places under the supervision of the 'sentry', although even in the wild foals are born during the daytime, as photographs taken in the Serengeti have shown. From this it seems clear that it is not the darkness but the feeling of security which may be the deciding factor of parturition or its delay.

On the basis of my own observations of free-range ponies, I can say with certainty that horses rarely look for sheltered places to have their foals and then only in very cold windy weather. If there is a feeling of security, the mares will foal when they are ready and will not delay the birth even during the daytime. In studs they seem to feel most secure after the evening feed, when the staff has gone home. Some owners like to keep a watchful eye on the mare and if, in spite of this, they are obliged occasionally to leave the box and thereby miss the birth, this shows that even their presence caused a certain disquiet.

On my own stud over a ten-year period on average almost half of the births occurred during daytime. Since this is contrary to the knowledge we have gained, it shows exactly how secure the horses feel and this, too, is apparent when horses of any age lie down in their boxes while people are about. This point was made particularly clear by the following occurrence. As an exception, one day a kind and devoted horseman, who was a stranger to the horses, took over the care of a mare due to have her fifth foal. Although there were all the signs that the birth was imminent, the night passed without this happening. When I returned home at midday, since the weather was so fine, I let the mare out – she cantered straight across the paddock with milk streaming from her teats. There, on the far side of the tree-lined paddock, she lay down in the open on a sunny patch of short grass and in about five minutes produced a lively, fully-developed colt foal. I could not help thinking that she had wanted to escape the worried gaze of my friend, and at last was able to foal in peace.

## THE BIRTH

The preliminaries to giving birth have already begun with the appearance of waxing and possibly milk, and these are also connected with the beginning of mother-love, which is often present even before the foal comes into the world. Mares usually remain standing during the early stages and may have strong labour pains. Then, generally, all Equidae lie down on their side, getting up and lying down several times, and, if everything is normal and the foal has turned to the right position – its upper line to the mare's spine and its forelegs stretched out – the 'water bag' will appear between the vulval lips. It should not be artificially broken. Then one of the foal's feet may appear inside the water bag and soon this will break, discharging up to two gallons of placental fluid. Normally, as I have said, all Equidae lie on their side, which is mechanically the easiest way of pushing the foal out. A foal weighs about 1/15th of his mother's weight and, if the mare is not lying on her side, this weight has to be pushed upwards through the mare's pelvis, which makes the act of giving birth more difficult. Weak or dead foals never turn into the normal position for birth, and the birth may be prolonged and difficult. I will not go into further details here.

A very large foaling-box is a necessity. This should be disinfected and bedded with fresh clean straw. During labour some mares throw themselves around and the feet of the foal would seem to be in danger, therefore one should try to keep the mare away from the walls of the box so that, when she does lie down, there is room for the foal to make its way safely. As soon as the head appears, the mare's labour pains are renewed, but the rest of the foal's body should follow in a short time. Because of the convenient formation of the pelvis of all Equidae and in addition, in comparison to other domestic animals, the strong muscles and good circulation, with a normal birth the period of delivery lasts from a few moments to less than half an hour. Then the mare generally gets to her feet, thereby breaking the umbilical cord at precisely the place where Nature intended. Outside interference should be reduced to the minimum, but it is of the greatest importance that

*Illustrations 67-77*

Plate 29   Early sexuality in foals.
*above*: Submissive facial expression on a Haflinger colt foal with a 'bad conscience', which had tried to mark the dung of a mare. His father Medicus corrects him by pushing him to one side with his head, but without threatening him.
*below*: Four-month-old Haflinger colt foal *flehm*s on catching the scent of a mare on-heat. In the background the stallion tolerates his harmless rival.

Plate 30   *above*: Those interested in behaviour should not be prejudiced by the horse's breed. The Rhineland-Belgian cold-blooded horse Steffl (*left*) and pure Arab foal Maymoona (*right*) watch the photographer with an intense expression.
*below*: Horses throw up their heads as danger draws near. The Lipizzaner mares Conversano, Bellamira and Traga come to attention.

(Continued page 119)

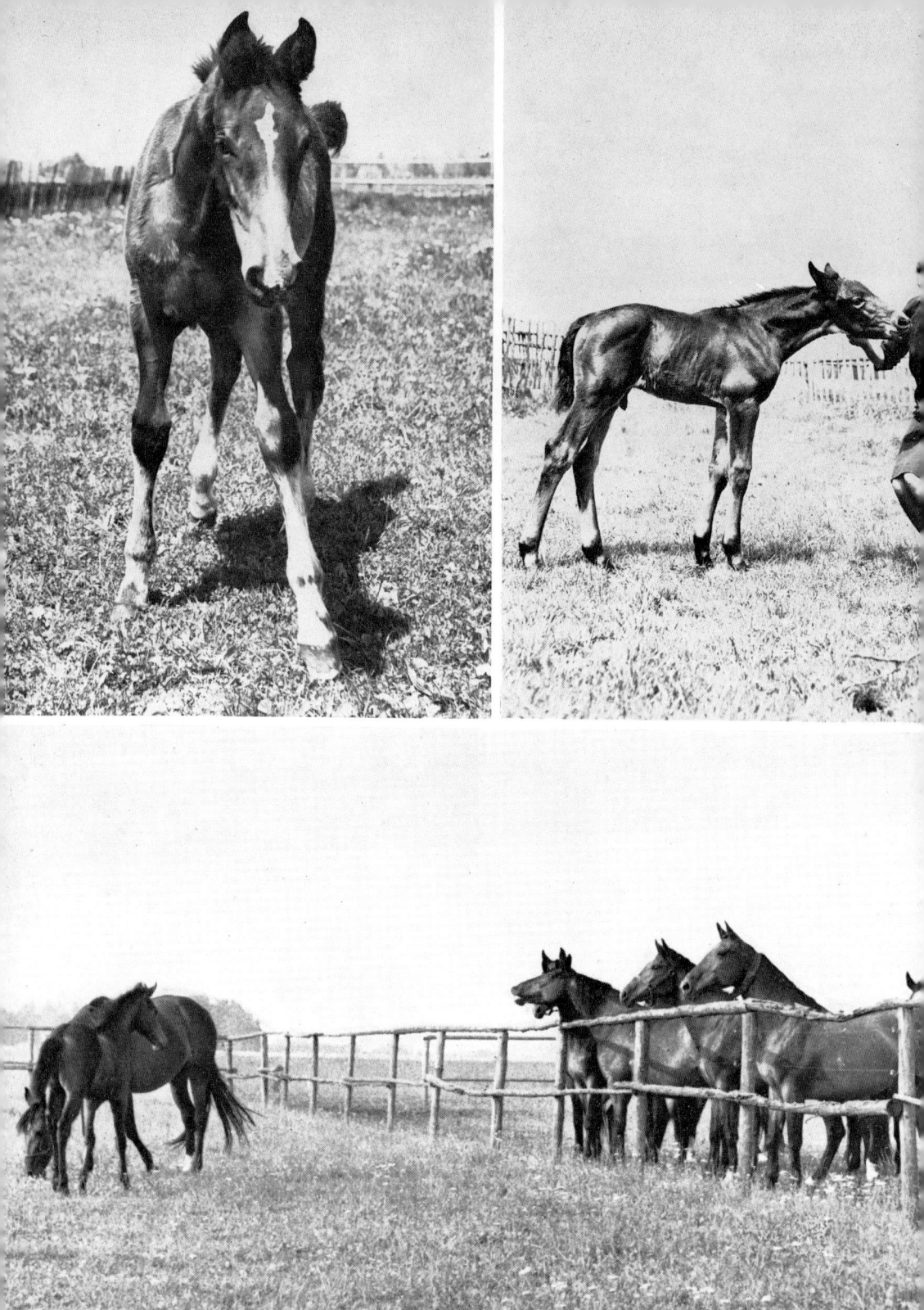

absolute cleanliness should be observed throughout the foaling period. It is obvious that the person who is caring for the mare should try to disturb her as little as possible. He should move slowly and talk to her quietly, and this will help a nervous mare, but as far as possible she should be left alone, so that she can concentrate on labour. To have more than one person present at a normal birth is unnecessary, just as whispering and creeping around either in the stables or outside from bush to bush on a range is extremely irritating for the mare, because the effect is that of a dangerous beast looking for its prey.

Equidae possess a fairly strong placenta which will tear during parturition, but which the foal must remove himself by shaking and sliding or by wriggling his legs. Strong healthy animals are capable of freeing most of their bodies and even their heads from the enclosing tissue, which is why a foal that has suffocated is generally weak even if it was not dead at birth. Mares that have produced several foals will lick and

Plate 31 *above left:* The Thoroughbred English stallion Marsilo XX wearing a rather distant expression.
*above right:* Mother and child wear almost the same 'muddled' expression – the ears show confusion. Andalusian mare and newborn foal.
*below:* Meeting of Hanoverian mare and Haflinger stallion. The mare threatens with ears laid well back, although the neutral expression of mouth and nostrils betrays the fact that curiosity is overcoming aggressiveness. The Haflinger is extremely interested and has an 'intense' look.

Plate 32 Assessment at close quarters.
*above left:* The warm-blooded foal Gobi hesitates as he approaches the photographer, who is sitting; the tail held high shows that he is ready to flee at a moment's notice.
*above right:* Like little children, foals enjoy licking things. Foal with the author's wife.
*below:* Two newcomers cause a commotion.

nibble their babies immediately they are on their feet, but this occupation, commonly described as 'licking dry', which is carried out at the beginning with great intensity, serves more for their smelling, and often tasting, orientation. As well as this first physical contact, there is vocal contact, which usually starts with a little whinny from the wet foal and is answered by the mother with a specific, very soft sound, which is heard only at this time. Licking and nibbling the foal's face and back last rather longer, but is not quite so thorough a process as that of carnivores; even the oft-quoted massaging effect of the mare's tongue to increase circulation is of less importance. The after-birth or 'cleansing', which in normal cases follows within approximately two hours of birth and causes some discomfort, is strong enough to make the mare lie down once more wearing an expression of pain on her face. Only in exceptional cases do horses eat the after-birth; with other animals this is not so exceptional, but it should be prevented.

## THE FOAL AND RECOGNITION OF IDENTITY

Apparently giving birth not only affects the mare but, judging by the tired and exhausted expression of the new-born, it is also a great physical effort for the foal, which certainly does not look at the world during the first moments of his life as cheerfully and happily as many writers would have us believe – possibly they have never been present at such an event. However, depending on its vitality, a foal will generally, within a few minutes, try to stand on its wobbly legs and will often overbalance, and tumble back to the ground. But, with an astonishing persistence, it makes repeated attempts until it actually stands completely still for a few seconds, with disorientated ears, as if it wants to make the most of its new-found balance. Mares which have never foaled before are sometimes afraid of the little stranger lying on the straw, but they very quickly lose their shyness when they see him standing on four legs like a proper horse.

At this point the urge for milk becomes so strong that the foal manages to co-ordinate his movements and takes his first steps to his mother's udder. On their unmanageable legs foals push and nudge their way carefully, unlike calves and especially lambs which punch and push rather violently with their heads at their mother's chest and stomach until, quite by accident, they eventually reach the right place. Generally the udder is full of milk and the foal has only to give it a gentle nudge for the milk to flow, then the foal gets a taste and more or less by accident finds a teat. The mare will help the foal by turning so that he can reach her udder easily and relieve the pressure by suckling. Some mares nibble the croup and dock of their sucking foals, this is usually answered by a hefty kick from a strong foal, causing him to tumble to the ground, so that the whole performance has to be repeated. If he manages to get his lips to the right place, she may try to help by pulling in her stomach and by a gentle, encouraging, quite special tone of voice.

From the rather persistent search between the mare's forelegs and perhaps chest, as well as between the hocks, it seems that foals have an in-born knowledge only of a roof-shaped angle but do not instinctively know the exact position of the udder. I would like to explain an evolutionary fact. The so-called pattern of the teats lies in the embryonic state in all mammals and humans in the angle between the forelegs and breast and the hindquarters, and, according to the species of animal, is variously developed. The 'pattern' is fully preserved in the dog and pig with their numerous teats; in others it is less developed, and in the elephant and monkey there are only two milk glands in the region of the breast, whereas with horses and many other suckling animals the teats are between the hindquarters. New-born foals always seem to search in the region of breast and stomach, the original seat of the numerous milk teats, although their target underwent an anatomical change long ago.

One of the biggest difficulties for our horses in finding the udder lies in the possibly false relationship between the height of mother and child. Domestic horses vary much more than wild Equidae, which is why foals, even with very similar forebears, can be born outside the average large or small. When the parents are of different sizes, the size and weight at birth of the offspring is governed more by the dam, but the foals of small mares bred by a rather larger stallion can have a bigger frame than is in accordance with the size of their body. The foals of zebras and wild asses are always born in proportion, and, for this reason, they always seem to be especially tiny with relatively short legs. They need only to keep their mouths, heads and necks at the natural suckling level, and to stretch a little upwards, in order to reach the teats without difficulty. Larger domestic foals with smaller mothers have far greater difficulty, because they have to hold their heads and mouths in a bent position to get at the udder, and, as this is an unnatural position, it may be several hours before they can suck successfully.

Hungry little foals have a typical suckling expression in which their tongues are folded like a V between their lips and a clear sucking noise can be heard. By extending its groove-like tongue, which is exactly the right shape for the oval teat, the foal gets his first taste of milk and he is then encouraged to suckle with renewed intensity, and his second effort is much more successful than the first.

I do not want to discuss the connections between hormones

and nervous stimulus or to give a detailed description of the anatomical and physiological results of the birth and secretion of milk. I shall only try to give an example of the strong link between the maternal instinct and the flow of milk. One of my warm-blooded mares foaled three weeks prematurely, without any change in her udder. She showed no feeling whatever towards the foal, which could not stand; she took no notice of it and even trod on its outstretched legs and body. It is most important that a new-born foal should drink the colostral milk, which is an antibody and acts like a mild purgative. The foal has to absorb this fairly quickly, so I massaged the mare's udder at half-hourly intervals to encourage the milk flow and fed the foal from a bottle. After about thirty hours and at least fifty udder massages, the milk began to flow and two hours later her teats were full. From this moment on the mare changed completely and behaved like any other mother, whinnying to the foal, licking and nibbling it and being very careful not to tread on it. The foal lifted its head to answer, but in spite of getting the colostrum and although part of the meconium had been passed, it became weaker and died thirty-six hours after birth.

The mare's maternal instinct was such that, once she had regularly given milk, she became quite wild when we tried to move the dead foal from the box, so we left it with her until she took no further notice of it. As soon as a foal appears to be lifeless and is not apparently breathing, the maternal instinct decreases, even if the udder is still full of milk. The mare appears not to notice the dead foal and may even stand on it. At the second attempt to remove the dead foal, the drama began again and we were once more forced to wait, until we thought of putting a rug over it so that she could not see the artificial 'movement' of the foal as we carried it out of the box; the mare then showed no further reaction. Later, when she was turned out in the paddock and had to pass her now uncovered baby, she smelt it for a moment and then, with no show of emotion, she went on her way. These facts are contrary to a perhaps sentimental notion, but are by no means unusual and apart from this, as far as concerns wild horses, these reactions are absolutely sensible, since horses on their own are in danger and have to follow their family group under all circumstances. Nature is kinder to animals in letting them forget faster than people are able to.

The next most important thing for a new-born foal after suckling is to ensure that the residual matter which accumulates in the rectum, called meconium, is passed. The bowels should be emptied within twenty-four hours of birth and the foal may have to strain to pass this meconium. However, if the foal has not cleared its bowels by the time it is approximately eight hours old, an enema of warm soapy water may be necessary. Under these circumstances it is advisable to call in expert attention.

The way a foal imitates its mother is a matter of importance in its life. Bird-watchers and ornithologists were the first to make the startling discovery that the ability to identify one's own species is definitely not inborn, as we may see in the farmyard when little geese or ducks waddle along behind a member of a strange species such as a hen. The identity of the species must be learnt during the first few hours through the mother-child relationship. A gosling imitates the creature or moving object which it sees as soon as it emerges from the shell and which has vocal contact with it, and Lorenz established that this reaction was caused by the first sight, movement and voice of the 'mother' object. Later it was discovered that mammals also have to receive an 'image' – although one cannot determine particular elements so exactly. In 1944, B. Grzimek, who was then an army veterinary surgeon, made his famous Kaspar-Hanser experiment with a foal which he took away from its mother at birth, before it had time to see her, and which he raised on a bottle. The foal looked on humans as its own kind, like Konrad Lorenz's goslings, and later was afraid of its own species.

Fledgelings and mammals always gain their sense of identity within a precise timespan depending on their species, which, as far as nestlings are concerned, must logically be shorter for those that leave the nest early than for dependent nestlings, which at first only see their parents. Bearing this in mind, it is sensible for wild Equidae and domestic horses in freedom to allocate a special position in the family hierarchy, even above the stallion, to a foal during the first two days of its life. The reason for this is that other animals may thus successfully be kept at a distance, so that the foal does not mistakenly accept another mare as its dam. Whilst the foal may behave like an automaton in its absolutely vital search for milk, it can clearly see forms and colours, can distinguish rough or smooth surfaces, and can differentiate between smells and sounds. I do not really know, however, which sense is most important to the foal in identifying its mother. The decisive phase in recognition comes in the hours immediately after suckling. I imagine that the act of drinking milk plays a large part in its reaction to images, since foals which have reacted strongly to impressions of colour or sound, say to blue jeans or my voice, at the beginning of their search for milk, still recognize these impressions but do not pay so much attention to them as they did at first. If one has had to help a foal to suckle, so that it thinks it must ask for 'help' and trustingly nudges one at every new effort before it is accompanied in the direction of the new udder, and if one returns to the box some two or three hours later, the foal will by then have recognized its mother and, according to temperament or disposition, will either take no further notice of one or will back shyly away.

I can pinpoint the moment of recognition almost exactly from observation of the birth of a Fjord pony foal. The foal was born at 4 am; by 4.20 am it could stand and ten minutes later it was suckling hard. Just after 5 am the older filly foal that I have already mentioned and which from the moment of the foal's birth had been drawn to it could refrain no longer from trying to make friends. At 5.20 am, when she finally became too friendly and started nipping the little one, she was pushed away by the mare. The exact period in which the foal learned to recognize its dam was between 4.30 am and 5.20 am, since, shortly after this, the family left its sleeping quarters and wandered off to graze, and the new-born foal without hesitation followed its mother instead of its new little friend.

It is not quite clear how mother and child recognize one another. Klingel suggests that zebra foals recognize their dam's special stripes, which in spite of the similarity of the animals themselves, varies amongst individuals. As far as foals are concerned, colour plays a major rôle: one of my foals, the offspring of a mare with a light-bay, 'metallic' coat, went galloping and neighing towards a newly-bought mare, which had the same metallic glint in her coat, although she was of quite different conformation to the mother. It was only the unfriendliness of the newcomer that made the foal realize its mistake, whereupon in great consternation it ran here and there amongst all the other brown horses, and between the two mares which had exactly the same colour coats, until it was able to recognize the difference in shape. Visual characteristics always play an extremely important part in recognition, and the sense of smell is important too. After branding and clipping the ponies which inhabit the Cantabrian Mountains all the year round, even the not over-sensitive Spaniards are careful how they free them.

Several men drag and push the mare and foal together, and hold them until both terrified animals recover from their fear, smell the other's noses and clearly recognize each other; only then are they freed and can gallop off.

The first few days of a foal's life are dominated by sleeping and drinking, whilst exercising its legs and getting to know its surroundings are of lesser importance. If foals are not actually resting, they suckle four or five times an hour and the frequency increases, so that they spend five to ten minutes suckling interspersed by periods of sleep, which last usually from half an hour to an hour, evenly spread over day and night. The time spent suckling depends upon the quantity of milk and lasts several minutes, during which a strong foal suckles both teats dry. The breeding qualities of the mare can be seen in her foal, which should be well-covered within a few weeks – a poor, thin foal with a big head and staring coat is a sign that the mare is not giving enough milk. It is not at all natural for mares with foals to be thin – an idea held by some rather mean owners – and the old country proverb should be better known: not only cows are milked through their mouths, but horses as well. Whilst they are suckling their foals, mares with a normal supply of milk cannot be given too much food, since most of it is passed on to the foal. On the other hand, the frequent suckling which stimulates the milk supply is unavoidable, though the milk usually dries up quickly when the foal is weaned. Like all other domestic animals, frequent suckling is good for the foal, as its small stomach is never overloaded and it suffers no digestive troubles. The persistent diarrhoea at the start and during the period of the first heat after foaling is physiological, because the milk has a different composition, and normally no treatment is required.

This is all of direct and special significance the instant a mare dies, and the little one has to be artificially fed. Although one cannot completely replace the immensely valuable colostral milk containing antibodies, the artificial mares' milk which is now available on the market allows a foal to develop into a useful horse the same size as his contemporaries, as long as one does not delay the feeding intervals to suit one's own convenience. However, to rear an orphan foal on a nurse mare needs a great deal of sympathetic understanding on the part of the breeder and can only be done if a very kind, motherly mare is available. This method often fails due to the contrariness of the mare and the orphan foal is easily distracted by a fresh source of milk – under natural conditions foals frequently try without success to suckle the stallion or other mares which have lost their foals. If they have lost their mother, even bigger foals look for a substitute – they generally find a way to attach themselves to another animal which allows a close contact and acts as a type of aunt. Of course the loss of a mother is not necessarily synonymous with death; ruthless early weaning or the sale of a foal has the same effect.

As the months pass, suckling decreases according to the foal's ability to graze or the availability of green fodder in the stable. If a great deal of grass or hay is eaten by the foal, it is an indication that the mare is not giving much milk, although, of course, the amount eaten does depend upon the type of foal. At an early age, too, it will try to share the mare's oats and bran, especially if this is fed in a container on the floor – a more natural way of feeding than the manger. This helps enormously when the time comes to wean the foal, which as far as domestic horses and ponies are concerned happens very abruptly at four to six months, depending on the breeder's ideas and the necessity to give the brood mare a rest, and a chance to recover for her new pregnancy. If the foals are not already feeding as suggested above and are suddenly deprived of their milk, they suffer great stress and become retarded.

However, so long as they are already feeding, milk supplements, which are available on the market, can be added to their feeds. Under natural circumstances weaning largely depends upon the length of the mare's pregnancy and, if the mare has too great a physical burden, the milk supply decreases, so that usually foals are weaned after about ten months. If the mare is not in-foal then the suckling relationship between mare and foal lasts much longer, sometimes up to one and a half years.

The longer weaning is delayed, the better it is for the preservation of the species. It is only because of this extra protein obtained from the mare's milk that the foals of the colder northern countries manage to survive their first winter without additional food and, even on a free range with extra fodder, the yearlings look in better condition at the end of the winter than do the two-year-olds. In the extremely barren southern regions of Spain, where drought is normal, Equidae mares appear to absorb or abort the foetus to avoid the physical demands of coping with both a live foal and an unborn embryo, and thus they usually bear a foal every second year, so that their foal can suckle for a longer period. On the extremely poor Spanish and Portuguese peasant farms, abortions through starvation used to be so common that nowadays, even in well-managed studs, mares are covered every other year.

Unlike some of the smaller mammals which remain helpless for days or even weeks, foals are able to follow their mothers for long distances very soon after birth. As soon as they have taken their first long drink of milk and rested, they exercise their muscles to find their balance and then try to co-ordinate their long wobbly legs. Depending on individual strength and vitality, these little foals very quickly develop the most astonishing ability to move. Not content with just walking round the mare, before they are even in a position to stand properly they canter round her with awkward little leaps, sometimes daring to try little bucks and jumps by turning on their hocks as part of the game. Healthy foals immediately practise different gaits like the amble, the tölt and the canter in three- and four-time, though their movements would not be passed by a dressage jury. At this early age they jump over all sorts of obstacles with more courage than they do later on, simply because they do not know the easy way or perhaps did not see it. They will try to scratch behind the ears, though they fall over doing so and getting up again is not very easy, if the forelegs have become crossed and have to be disentangled into a parallel position. The more the newborn foals gather strength and the longer their games last, the faster are the circles made around the mother. Like little children, their curiosity and tireless efforts to copy or mimic are the chief ways of learning. As they venture further afield, they will lick and nibble at the most unlikely objects, and try to 'feel' them with their forefeet, and in this way copy their mothers for the first time. Since their necks are still very short, they have to do the splits like giraffes and bend their legs in other awkward ways in order to nibble at grasses, which are then usually pulled out roots and all.

Stabled mares are sometimes so careful of their foals that they can be a danger to their fellow horses or people, threatening to bite and kick them. These aggressive mares can rarely be calmed by the voice, and steps have to be taken right at the beginning to see that one's own dominance is retained, if one does not want to have difficulty in handling them for a long time afterwards. This kind of 'vice' in domestic horses is basically due to the structure of the hierarchy, which allows even the lowest ranks to keep other members of the herd at a distance during a foal's early days. On a free range the mare will still prevent her new-born foal having contact with strangers and the same thing happens in a stud. Characteristically, in my own stud, the behaviour of the home-bred mares and that of the adult mares which have been bought in are quite different towards me, the latter are usually much more aggressive and their threatening attitude lasts quite a long time, and there is little one can do about it. After giving birth, some horses are not only tolerant towards their own species and people, but they appear to bother far less with their offspring. However, it is not clear how much this is due to and is linked with individual physical or possibly hormonal disturbances and even with neurosis.

The very close relationship between mother and child in the lives of wild Equidae can be interrupted by external exigencies; thus Grévy zebras leave their new-born foals in a form of general nursery, since the little ones cannot follow their mothers on the long marches to drinking places. They remain hidden in the long grass like fawns and wait until their mothers reappear. Onagers, too, are said to leave their very young foals for several hours. Some years ago Ebhardt thought that the same thing might have happened with the ancestors of the Arab horse, and it does seem possible, since both Arabians and very well-bred Thoroughbred foals sometimes have extremely weak fetlock joints during their early days, often actually walking on the joints, so that if they were forced to go on long marches they would soon hurt themselves. From my own observations such foals never appear to be as well-developed as the sturdier warm-blooded foals and especially the foals of the robust 'cob'-types, which have exceptionally short pasterns and in all their proportions appear to be more mature than the extremely slender-boned foals of eastern ancestry. Since onagers, too, specialize in fast action and also have remarkably weak fetlocks, an analogy between these eastern horses of similar build and the behaviour of their new-born foals is more than possible. The conformation of pony foals is similar to that of the Grant zebras and they, too, as far as we know, do not leave their foals alone.

# 6

## GROWING UP AND PLAY BEHAVIOUR

Human children and young foals have two things in common: the urge to play games and curiosity combined with a greater capacity for learning than at any time in their later life. In addition, foals have to be able to satisfy their elementary needs in order to grow into a mature horse with a harmonious physical and spiritual balance.

### INCORPORATION INTO THE SOCIAL STRUCTURE

After the first days of learning to recognize its mother, the foal normally comes into contact with the other members of the group and gradually practises the repertoire of his inherited social behaviour. Although at the beginning the mares take great care of their foals and protect them against the family, strangers and others, they are not spared during their childhood what our sociologists like to call frustrations. Normally, of course, horses get on fairly well together, even in the herds where there is no stallion, and they do not bully or injure animals smaller than themselves – the ranking order, which always brackets the foal with its mother, prevents this to a large extent – even so, mares are not usually very gentle with other mares' foals and they do tend to threaten them, and may resort to biting or kicking, if the foals get in the way. Sometimes the old top-ranking mares guard their positions jealously, although the foals' submissive behaviour helps to prevent excesses. Extreme reactions of this kind, which I have observed when the mares were actually viciously chasing strange foals, are, I consider, a faulty reaction caused by an unnatural way of keeping horses and I have never witnessed this in herds where a stallion was present.

## RUNNING, CATCHING AND FIGHTING GAMES

Playing is a typical behaviour pattern in all young animals. After the foal has spent the first few days galloping wildly around his mother and exercising his limbs, he looks for playmates. As one might expect, young horses spend a great deal of time playing running games that last several minutes, in which they race round one behind the other and which seem to develop into 'catch-as-catch-can'. Two play at this game,

*Illustrations 78-89*

Plate 33   Assessment by smell both close and distant.
    *top:* Nubian wild ass mare getting wind of a scent in Hellabrunn Zoo.
    *centre:* The Andalusian stallion Yesquero also gets wind of distant friends.
    *bottom:* The Holstein half-bred stallion Landgraf II sniffs closely a mare's urine.
Plate 34   *above left:* Horses, too, show their tiredness by yawning. Bosnian stallion Miško XII.
    *above right:* During their first hours foals can *flehm* and this foal is only thirty minutes old.
    *below:* This northern Spanish pony stallion *flehm*s flanked by two in-season mares. On the left, near the old mare, is her yearling.

(Continued page 135)

seldom more and there is a faint similarity to the behaviour of mutual herbivorous and carnivorous ancestors, since one of the foals always seems to try to take short-cuts, and this is answered by kicking and a faster tempo on the part of the playmate. This form of 'catch-as-catch-can' game certainly reminds one of an exercise that is very necessary for young animals in the wild, which helps to prepare them for a possible emergency. However, conscious training must not be evaluated too highly, since the wish to play is not tied to necessity, but appears both for humans and animals to be the beginning of every intelligent interest; it is quite another matter that horses remain at the same stage as what is, for us, a phase of early childhood.

The intention to break into a playful gallop, alone or in the hope of attracting others, is shown by lifting the tail very high. A foal's tail has very short hairs and looks rather like the upright docked tail of a fox terrier. This visual signal, which is the same for adult animals, even heavily in-foal mares which because of their condition have very little inclination to gallop, alerts every horse in the neighbourhood that is ready for a

Plate 35 *above:* Dozing faces with characteristically relaxed ears and hanging lower lip.
*below left:* Fjord pony foal in a deep sleep.
*below right:* One has to get used to sudden light. Grey Kladrub foal blinking in the sun.

Plate 36 Strangers greeting one another.
*top:* The Trotter mare Onda being inspected. The aggressive and higher-ranking Holstein mare Aischa (left) with her neck bent to show off. The lower-ranking mares have to wait for their greeting.
*centre:* Two-year-old Fjord colt in a friendly greeting with the strange new mare, which reacts in a slightly dismissive manner.
*bottom:* Yearling mare greeting a strange Fjord gelding with an 'archaic look' and showing only the lower incisors; the ears of both animals are pricked in a friendly way.

gallop, even if the signal is partly obliterated by a heavy growth of hair or, depending on the breed, varies from horizontal to curling over the back. Recognition of this signal of wanting to gallop, plus the typical snorting which accompanies it, which is also a warning of something strange, can be useful to anyone studying horses at large and a shock can be avoided when the herd suddenly breaks out into a gallop. If weather conditions are good and if they have enough oats inside them, even older horses of both sexes enjoy a good gallop, which soon becomes a wild race with bucking, kicking and little squeals of pleasure. Although a visitor on his feet can enjoy such a spectacle, the position of a rider is less happy, especially as it frequently happens after school horses have spent a day in their boxes. A strong and energetic urge to gallop is very infectious and even elderly equine grandmothers put their tired joints into action and try to keep up with the wild onslaught. This has absolutely nothing in common with the panic flight, which I shall later describe.

Very early in life the play behaviour of colts differs noticeably from that of mares or fillies. All the activities of filly foals are, with the exception of galloping, much more seemly than are those of colt foals, which quite soon seek out companions of their own age and sex, because they are more interested in games than in fillies. Contact games with the filly foals, nibbling each other and continuing to more intensive mutual coat care, are soon turned by the colt foals into rough play. They will pull and push one another about and gradually the first steps are taken towards the future fighting game. In contrast to coat nibbling, when foals, like their seniors, approach each other diagonally and the expression on their face clearly shows their intentions, colt foals begin their scuffles by approaching head on. At first this is shown by an awkward semi-rear, later expertly demonstrated, and is followed by an energetic chase during which one colt tries to 'catch' the other by biting his neck or shoulders. The friendly intentions of this game are always revealed by a rather cheeky expression with the ears pointed forwards. As we shall see from studying the identical preliminaries to fighting behaviour, even the actual fighting game begins with the same impudent facial expression since even strong scuffling foals know perfectly well that these are only games and can be broken off when one of the contestants shows that he has had enough by kicking out with both hind-legs.

If no other foals of the same age are available – and this certainly leaves unfortunate physical problems of development in its wake – then the dam receives a challenge to play. The foal bumps into her, disturbs her grazing by dancing close to her and, in fact, uses every way to get her to play with him. We have already mentioned that foals derive a great deal of fun from seeing their mothers rolling. If the mares remain on the ground for a while then their offspring often try to jump on their necks or backs. Colt foals especially try to jump their mothers in play. I don't think this has anything to do with training for their rôle as stallions later on. Although many mares co-operate good-naturedly and play with their colt foals by cantering with them, the latter do not really enjoy girlish games, so they look for substitute playmates amongst other animals such as dogs, goats, sheep and even humans. If they have the choice, they prefer playmates like themselves such as a donkey foal, a calf or a dog rather than their human friends.

## GAMES WITH OTHER ANIMALS AND HUMANS

When foals play with animals other than their own kind, it is fairly plain that the two do not always understand each other's language and mistake the signs and signals. We own a large, kind and very strong Leonberger bitch which loves not only children but especially small foals. For want of other dogs to play with, she cannot think of anything better than to romp with the foals out at grass, although generally she can only cope with them for a short time. The reciprocal way of greeting to the head, flank and hindquarters has a similar pattern, so the introduction is fairly successful and the foal soon begins to nibble at his partner's coat. Her friendly licking of the foal's mouth is frequently misunderstood by him, as he regards it as a challenge to play at biting and therefore lays back his ears in hostility, which the dog fails to understand: she is really deeply hurt when the other bites too hard or goes for her with his forefeet. The importance of man's superiority over his horses may be seen from the effect of the terrified dog giving cries of pain and beating a hasty retreat, whereby she loses the respect of her inferior and is chased for some distance. If the dog then starts another fight and wins, she could cause the colt's feeling of superiority to disappear, however this is just as unlikely to happen as it would unfortunately be with many inexperienced horse-lovers. If a human occupies the position in default of playmates of the foal's own species, he will soon find, even with tiny foals of the smaller breeds, that their pushing and pinching are more than he can bear, and then he becomes afraid, whereupon animals that are naturally difficult or aggressive can be spoilt for the rest of their lives. A violent reaction caused by fear on the part of the human playmate will be regarded as unfair and will not be understood by the foal, and this once more affects the harmonious relationship between horse and man. On principle, no individual should be reduced to the position of a plaything either for future show jumpers, riding school horses or funny little Shetland ponies, but it causes considerable problems in his association with the animals. There is nothing more distressing than for the owner of horses to have to conceal his insecurity with his own animals, which are always supposed to be good in front of visitors. We need to possess a great deal of sympathy if we want to be able to play with horses, and, as I wrote at the beginning of this book, we must be able to understand their language for the game to be enjoyed by both parties. Even with the greatest confidence on both sides, in every situation the human must be the highest ranking partner. This is shown by his quietness, his movements and voice, and this prevents a stallion which may be rearing in play from turning it into a fighting game or even a serious fight. The preservation of the right distance, which can produce respect, flight or attack, rests on considerable experience: some circus trainers are past masters of this art of control.

## EDUCATION FROM WILD TO DOMESTIC HORSE

Since the ability of young foals to learn is considerably greater than that of older animals, some people think it is a good idea to begin their education fairly early. I am not, personally, entirely in favour of this idea, since I think that a foal should develop its normal behaviour in the company of its mother and other horses before people begin to confront it with a variety of rather unnatural things. Thus, with my own foals, I pay great attention to gaining their confidence, as with the rest of the herd, before they are actually weaned and before we begin any form of training, and I avoid all situations which might bring out feelings of aggression. Even during the first few hours after birth, individual differences are noticeable. Apart from extremely shy foals, which are afraid of being touched, there are bold, friendly and even short-tempered new-born foals, which, before they are even dry, put up a defence or want to be aggressive. However, in spite of generations of domestication, all foals are born naturally wild and, since training means that we overcome their fear of us – after all, a tame animal allows itself to be touched and handled by man – every new-born foal has to be trained and taught confidence. Of course their reactions have been conditioned and even changed by domestication in contrast to genuine wild Equidae, but, even so, the inborn tendency to run away is always present to some extent. The influence of the mare on the process of being tamed cannot be overemphasized and her behaviour sets an absolute example, which is one of the reasons why the nervous or quick-tempered mare should be tied up when one handles her foal. It is well known that temperament and character are inherited to a far greater extent than exterior characteristics, so that for domestic purposes, under the saddle and in harness, horses have as a matter of course been selectively bred for character. Over several generations quick-temper can be influenced by patience, etc., but, since the ability to win in competitions is often coupled with a difficult temperament, selective breeding is made more problematic.

Without going any further into the Lorenz school of reasoning on inherited aggressiveness and its stormy rebuttal by many sociologists, I would like to make absolutely clear that there are horses which, from the moment they are born, are very ready to be quick-tempered and aggressive, and that they react promptly and energetically to any disagreeable stimulus. The general opinion of many real animal-lovers, that horses are born good and are spoilt by bad handling even to the point of being ruined, is only partially correct. One must not misjudge absolutely correct behaviour on the part of animals in the wild simply because such behaviour is dangerous for the human concerned. Horses of different breeds and especially of different types probably also possess different degrees of aloofness or of 'individual critical distance', and they will become aggressive if this is disregarded. A strongly felt 'individual critical distance', which logically makes association with the animal more difficult, is in fact not always coupled with a more marked desire to attack. There are animals which will attack at once, whilst others try to get away by fleeing until the critical distance is so reduced that they, in turn, reverse their position to attack. This latter situation is very dangerous, because it has developed out of great fear. However, I believe that even these horses, in spite of all their individual idiosyncrasies, can be trained to be useful animals.

Colt foals have a far greater inclination to play than filly foals and this continues until they are fully grown – rather like their human counterparts – and they can possess a remarkably early sexuality. I noticed this particularly with some Haflingers when the family stallion was running with his herd in the same paddock. Both the three- and four-month-old colts showed an astoundingly complete repertoire of sexual behaviour, ranging from an examination of a mare's droppings to an attempt to cover a mare in season. I do not know whether they were simply copying the example of the stallion and it was a matter of seeing and learning, or if it was a normal reaction of the Haflinger, which matures very early in any case, such as would have occurred without an example set by the stallion.

# 7

# COMBAT BEHAVIOUR AND FLIGHT

As the colt foals grow older, their character as entires manifests itself more clearly and the fighting element comes to the fore, until it includes all mannerisms of behaviour and collective tactics which can be used in time of need. Although the speed of attack and reaction are, at first, very slow, they increase gradually.

## FIGHTING GAMES

In contrast to a real fight with its unmistakably threatening expression and bearing, young entires ready for a sparring match wear a friendly expression. Before a real quarrel starts, they move towards each other and show off with head high, neck flexed and doing a form of collected trot, which, as they get close to each other, often becomes a cadenced passage. They circle and mutually smell each other, and stamp the ground with a forefoot; then suddenly, with loud 'war cries', the signal for battle is given. To prevent injury there are normally only two sparring partners and in the separate phases of the game this is adhered to. If a stallion opens the battle by trying to force his opponent down on his knees with his neck, the other reacts by using his neck to ward him off or else tries to swing sideways and by this means to force the rival to the ground.

Even if they manage to avoid this and, after a preliminary show of intention start biting the muzzle, and then attacking the forelegs, shoulder and ribs, both stallions fight equally and each one defends himself by pulling his leg away or by kneeling on his fore-fetlock joint, whereupon the attacker also kneels. This method of combat is usually interrupted by a short whirling round from head to tail in an effort to snap like lightning and to avoid being bitten on the hindquarters. In these sparring matches the youngsters occasionally kick each other by dodging sideways.

## THE STALLION'S METHOD OF FIGHTING

The circling action and mutual attack with the teeth can become so deadly earnest among the Grant zebras that the opponents even adopt a sitting position like dogs, sliding around on their hindquarters to continue the fight. However, this aggressive sitting position is used only by the Grant zebras and up to the present time has never been observed either in other wild Equidae or in domestic horses. In serious fights most Equidae begin without much previous skirmishing: they pretend to make biting attacks or rear and fight with the forefeet, and they bellow loudly. Neighing is scarcely the right word to describe these angry, noisy shouts. They attack each other at once, standing on their hindlegs and trying to attack with the forefeet or with lightning bites aimed at the neck or crest of their opponent; yet even when they fight in real anger, both rivals always fight on equal terms. As rearing is a tiring exercise, even for well-muscled and strong stallions, both rivals return to all four feet, circling and snapping, until they are ready again to rear up for the very effective hitting and biting combat. If one of them succumbs, he usually takes flight, defending himself by kicking for the short distance during which his attacker follows him. Ebhardt apparently observed a battle between Icelandic ponies, when the defeated stallion uttered a throat-rattle perceptible only to his opponent to admit defeat, whereupon the attacker stopped and allowed the other to flee.

There is something extraordinarily dramatic about such a short, sharp, intense battle between stallions, not least because of the vocal accompaniment underlining the truly fascinating spectacle, which is one reason why it is a favourite theme in novels and films about horses, and inspires authors, scriptwriters and behavioural researchers to write colourful scenes and very imaginative descriptions. Serious studies in behaviour rarely describe genuine fights under natural circumstances and then only between fully-grown stallions when an in-season mare is about to be stolen from the family group. According to Klingel, the removal of a filly from the group is usually settled without a fight and is achieved 'simply by chasing off, without the opponents coming into closer contact'. This is contradicted by numerous descriptions of other wild Equidae, especially of the Przevalsky horse and the Asiatic onager. Careful examination of these descriptions leaves one in doubt as to whether most are not in fact about animals in varying degrees of captivity or even the few remaining free animals inhabiting an enclosed area and not about animals that are entirely undisturbed. Russian sources indicate that the Przevalsky horses and the different onager, kiang and kulan strains behave in a much more aggressive manner and fight so furiously that the animals are hurt or even killed.

It appears to me that a battle between rivals to the point of the destruction of an opponent is not very likely, since under natural circumstances it does not occur amongst the other higher vertebrates and is not designed to help the preservation of the species. Its much-quoted value as a means of selection – only the strongest male is allowed to mate and to pass on his characteristics – has exactly the same effect if the inferior is driven off and continues to exist as a biological reserve. Most of the furious fights described took place in enclosed areas, which were probably too small for Equidae requiring an extensive 'individual distance', so that the winner was not able to drive the loser far enough away and, since the latter then had to remain in close contact, the fight would have to be continued and either horse might be injured or killed. Investigations are being made into the behaviour of independent wild Equidae, and we may then be better informed on the differences in combat conduct of individual Equidae.

As far as domestic horses are concerned there are differences between breeds and types in their aggressiveness and also differences in the intensity and speed of the stallion combat. Horses with oriental blood seem to be particularly impetuous and manoeuvrable, and this would appear to be substantiated by accounts of their superiority in fighting much bigger opponents. Their remarkable ability to bite, which I have noticed, causes horses of cold-blooded ancestry and slower reactions to retire willingly from the battle.

As long as stallions are psychologically equal (size does not really matter here) and are kept under natural conditions, contestants are rarely injured because, as in other spheres, there are rules of fair combat which are always observed. Apart from genuine accidents, stallions are only, in my opinion, seriously injured when they so differ in exterior and interior characteristics that their rules of fighting are no longer appropriate, and thus the quicker of the two confuses and catches the other unawares.

Briefly here, let it be noted that circus dressage, like the High School movements, is based on these elements of stallion combat: the passage and piaffe are taken from the stallion's desire to show off, whilst the circus rearing-act and the levade, croupade and capriole originate in the stallion's rearing position in battle. Since most school figures, in circus riding as well as free dressage, classical High School and higher dressage, deal with the numerous movements which belong to the typical behaviour of a stallion, in my opinion it is in a way very bad taste to apply such exercises to mares, as is the custom at the present time.

## THE MARE'S METHOD OF FIGHTING

Stallions only occasionally use their hindlegs in battle, since kicking is usually the defensive action of a mare and is seldom meant as an attack. Of course, mares do react in a masculine way to other enemies such as carnivores threatening their foals, in which case they fight them off by striking with the forefeet and biting, but their battles for superiority, which are generally the reason for fighting, are decided by kicking. In such instances, they only hurt each other if one of the rivals refuses to back down in time or if two strange, very high-ranking mares meet and fight an unusually severe and lengthy battle. On the whole, however, the threats of a strong personality are sufficient to cause the other to give way. When mares not in-season are bothered too much by the stallion, they will, as I have said, kick out, sometimes ejecting urine. This combined reaction may serve to throw the urine with the hooves into the face of the enemy, similar to the repulsing action of other mammals, like the well-known squirting of the skunk.

From many accounts, we know that mares will defend their foals against wild beasts by forming a circle round the foals, with their own hindquarters facing outwards at the ready. It is also reported of wild stallions that they take their guard duties very seriously and energetically drive off the preying animals by striking with the forefeet, sometimes killing them. It is understandable that such methods of attack or defence are more moderate in the case of mares with foals towards dogs, which, like the wolves of earlier times, are followed and chased sometimes for quite a distance out of the paddock by especially militant brood mares and by stallions using their forefeet and teeth. Striking with the forefeet is, in fact, an older evolutionary method of fighting than kicking with the hindfeet, which is phylogenetically much more recent and is a speciality of the various species of Equidae. We humans fight naturally in the original traditional method like two entires battling on their hindlegs, since we too, if we have not been trained in unusual techniques, fight our physical battles with our fists. Using one's feet on one's opponent is regarded in the West as exceptionally unfair, and the Asiatic methods of combat which are now being taught, with their cunning blows and special kicks, are rather strange phenomena to the average European. It is a matter of opinion whether these ways of fighting, ranging from the boxing in Thailand to Japanese karate, are a phylogenetic evolution of human combat behaviour, which has just the same confusing effect as the lightning bites of an Arab on a cold-blooded horse.

## THE CRITICAL DISTANCE

Besides fighting amongst themselves, the aggressiveness of domestic horses towards other species is particularly significant, since humans are often personally involved. Usually we are responsible, because we have infringed the individual's critical distance, perhaps partly by inexperience, so that it may not be a question of genuine aggressiveness but a defence against the supposed enemy – man. In this case, horses fight with their own typical offensive and defensive weapons, i.e. mostly with the forefeet, teeth and hindfeet, which they would use against one of their own kind. Since we are so much inferior, we have to resort to using physical strategy and a form of hypnosis over a much stronger animal in order to dominate him. We have used the expression 'critical distance', which was postulated by the Swiss zoologist, Hediger, and means the point at which a given distance between the animal and the advancing subject has been so reduced that the former will not take flight but will attack. This is defensive aggression and is practised by all animals in the manner typical to the species; horses of both sexes kick with their hind feet. If a dog is cornered, it will bite out of fear, but a horse in the same situation turns into a frightened kicker. In contrast to aggressive fighting conduct, which is also denoted by striking with the forefeet and biting, fearful or frightened stallions and mares turn their hindquarters to the source of danger, as their hindfeet are their strongest weapon and can do the most damage. On the other hand a member of the cattle tribe, whether born polled or later de-horned, uses its head for resistance if it finds itself in a similar situation.

## THE QUADRUPED TEST

Some authors describe Equidae as animals highly adapted to flight, which not only means that they are the fastest of all hoofed animals but is a high-sounding expression to describe them as ever ready to flee. I think it is wrong to describe horses as animals of 'flight', since their readiness to flee is nothing like so great as that of some other herbivores, and wild Equidae readily defend themselves against quite vicious carnivores. That is why we are told of Siegfried in the Nibelungen Saga that 'he fought and conquered the four aurochs and the terrifying stallion'. The horse did not flee! Domestic horses dominate cattle, which they chase by biting and kicking them as they please – and normally cattle keep a respectful distance. As far as bull-fighting with horses is concerned, when the bull tries to horn the horse, there is, apart from the bull's increased aggressive drive for which he is selectively bred, an exceptional situation in as much as the bull has been so greatly irritated by the journey and the earlier imprisonment in a dark pen that he no longer reacts in the normal way. On the range he rarely attacks mounted cowmen, in contrast to pedestrians. Ignorant animal-lovers are prone to think that horse races, too, are nothing more than organized flight, in which the animal most terrified of the whip runs fractionally faster so that he wins. In spite of this impressive theory, all the experts reject it, since well-fed healthy horses enjoy a good gallop and often run races against each other in the paddock, in which each tries to get to the front.

Brood mares with foals, as we have seen, form a circle with their hindquarters outwards to protect their foals from wolves, etc; this contradicts the 'animals of flight' theory, as does the further observation that zebra mares and stallions will attack hyenas with their forefeet, and, although the latter possess little courage, they do have powerful teeth. The occasional playful but often hostile pursuit of a dog out of a paddock serves to strengthen the argument. This is a traditional memory of his arch-enemy the wolf, since in the prehistoric habitat of the wild horse, the wolf, like the Stone Age hunter, was a most dangerous enemy and in severe winter claimed many more victims than all the other predators combined. Domestic horses still retain this traditional memory and also retain the will to flee.

The well-known scientist Zeeb was able to produce a startling effect by imitating a wolf in his so-called Quadruped Test. He went down on all fours like a dog and moved around in view of different groups of horses. His appearance caused the horses to come to attention; they watched this extraordinary apparition with all their senses alert and took to flight with warning noises as soon as it approached. If he remained motionless, the horses would approach him, heads held high and ears pricked, so it is plain that he awoke their collective curiosity. However, the moment he moved, the effect was as I have described above and the horses turned and fled with startled cries, until they had put a considerable distance between themselves and him. He was recognized as a human only when he got to his feet, at which point the animals changed their position and, regaining confidence, would approach again. It seems that all horses of various degrees of domestication react in the above manner with the exception of Shetland ponies and here an hereditary factor could be involved through the lack of the arch-enemy (wolf)

*Illustrations 90-103*

Plate 37  Neighing to make contact.
*above:* The Haflinger stallion Nasir in a light piaffe begins to call to his mares.
*below left:* Nasir whinnies loudly; only the lower incisors are visible.
*below right:* Trakehner stallion Komet neighs with a slightly aggressive undertone; the upper incisors are exposed.

Plate 38  'Archaic greeting expression' when showing off.
*above left:* The two-and-a-half-year-old, Noriker, spotted stallion Pipp-Elmar has no tushes to show, as he is still too young.
*above right:* The French Trotter stallion Quel Espoir C uncovers his well-grown tushes.
*below right:* The tushes of the Polish Arab stallion Wisznu cannot be seen because of the length of the diastema.
*below left:* The angle of the neck and head, and the evenly opened mouth of the Thoroughbred stallion Ladykiller show the friendly way of neighing to make contact.

(Continued page 151)

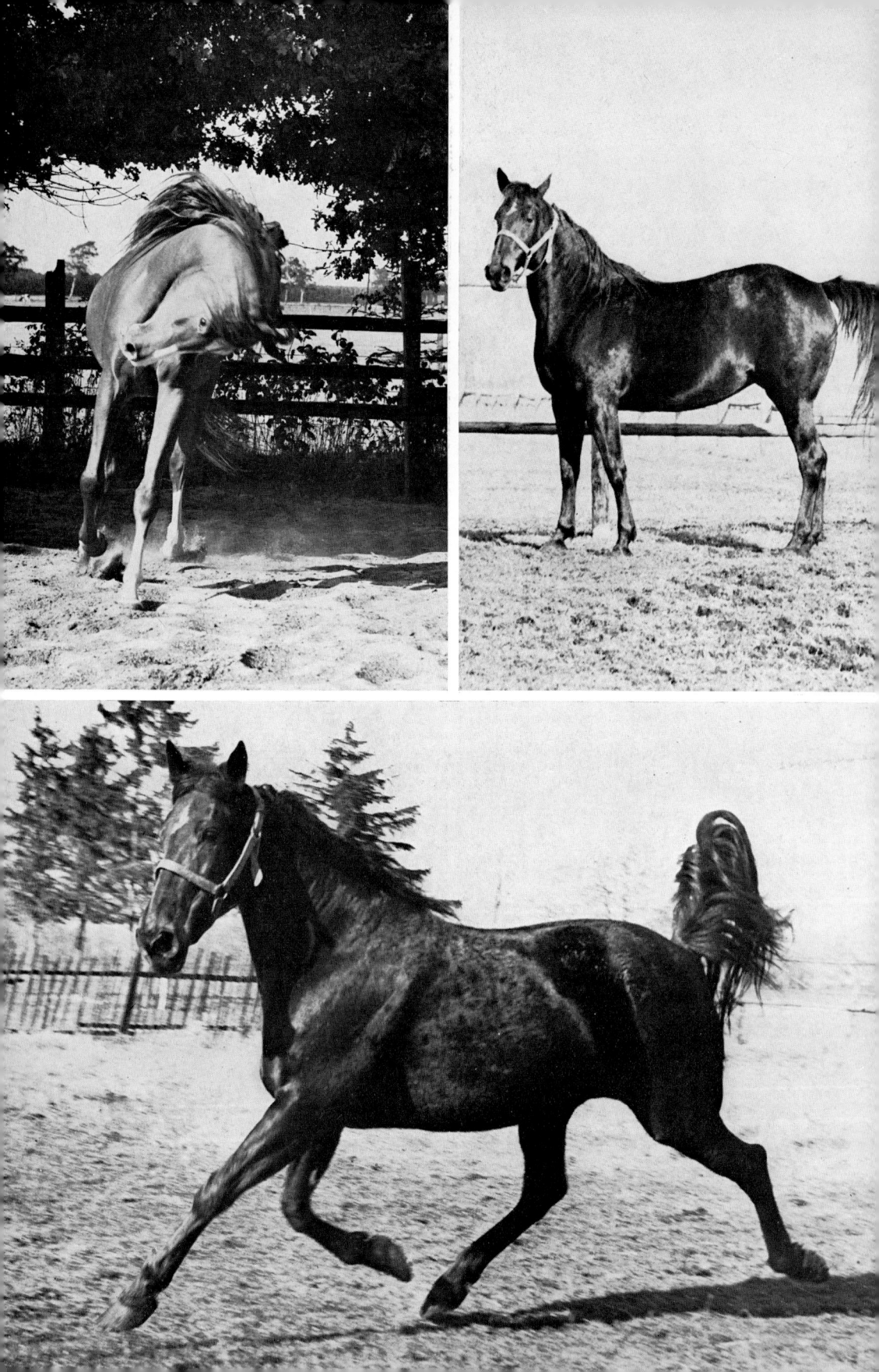

on the Shetland Isles, as the ponies remain almost totally unimpressed by this four-footed apparition, no matter whether they are kept in large reserves or in zoological enclosures.

Apart from the wolf image, some horses appear to carry a feeling for danger from above, from trees or rocks, since they panic if a shadow falls in front of them. Possibly this inborn fear arises from the time when an attacking leopard sprang from above, since the latter's habitat probably overlapped the most southerly area of the original wild horses.

## THE FLIGHT DISTANCE

The distance between a real or possible danger and the horse is best described as the 'flight distance' and it must not be transgressed if the animal is not to flee. It is presumably greater or smaller depending on the innate recognition of an enemy or

Plate 39   Showing off.
*above left*: An impressive expression by the unforgettable original Arab stallion Gazal.
*above right*: Holstein stallion Landgraf II tries with a dancing step to impress a mare.
*below left*: The most impressive action of a stallion is the piaffe. The Spanish Rejoneador Bohorque mounted on his magnificent Andalusian stallion at the piaffe.
*below right*: The pure-bred Arab stallion Darius in rhythmic piaffe with correctly placed hindquarters.

Plate 40   *above left*: Shaking movement of the head to show off and a sign of aggressiveness which cannot be put to good use – Samun, pure-bred Arab stallion.
*above right*: Lifting the tail whilst standing still signifies the intention to break into a canter; the expression on the face, however, shows uncertainty.
*below*: The trotter mare Onda's facial expression and tail signal are in accord.

the frightened behaviour of other members of the herd. It can be completely reduced by familiarity with a potential enemy like a man or dog and, in doing so, other differences relating to space can be seen. Horses that have no chance of escape, and those of friendly disposition or those that have been tamed in the sense of having been broken in by a cowboy or gaucho, completely stifle their instinct to flee, which means they allow themselves to be caught, bridled, saddled, etc. Stabled horses, of course, completely forgo their natural aloofness or 'flight distance'. When turned out in a paddock or in absolute freedom, many horses that have no particularly friendly relationship with people replace this 'flight distance' by a refusal to be caught, which varies according to the horse concerned and his confidence or lack of it. This distance between himself and his potential enemy is usually maintained at the walk and, although he does not actually run away, he shows a moderate form of flight, similar to that of wild Equidae facing carnivores that are obviously not hunting.

## FLIGHT AND PANIC

All Equidae have the remarkable ability to observe the whole of their natural surroundings and they register substantially more than is commonly supposed. Thus zebras are perfectly aware when a pride of lions has fed, since they then reduce their flight distance to one of avoidance or, alternatively, if the pride is hunting, the distance is increased according to the degree of danger and results in a real flight at the walk or gallop. As long as the retreat is under control, the animals flee in the normal marching order which they have when they roam, led by the highest-ranking mare with her foal and then her yearling, followed by the other mares in the same formation, whilst the stallion defends the rear. It is only when they panic that they flee wildly, without any consideration of family or individual differences, sometimes increasing in numbers to an enormous herd. As soon as the panic is over, this mass divides into the smaller loosely-connected groups. In the different species of Equidae panic does not occur with the same frequency or for the same reasons. The ass, a true mountain animal, never panics in flight, as it might easily have the unfortunate results of causing him to pitch over a ledge to his death. Animals which inhabit the steppes or open country, however, get into a state of panic much quicker and gallop away from the threatened danger, and thus the species is preserved.

Some well-bred horses seem to lose their nerve; they take fright at apparently harmless things which may perhaps be unusual or strange, and they subsequently run away without seeing any of the obstacles in their path, thus causing themselves even more harm. As animals of the steppe, the flight is obviously sensible, but our fenced-in, man-made world creates danger. Luckily even the fleetest of our domesticated horses rarely reacts in this manner. Animals and people only start to panic out of fear of an inexplicable danger, which then takes on exaggerated proportions causing irrational reactions. By contrast, in an organized flight, the fleeing Equidae have estimated the danger and the behaviour of the enemy, which is why they flee in their order of precedence until they are well out of the danger zone. Under natural circumstances, they will not tire as they do when panic overtakes them.

Horses can become frightened by unusual sounds, but usually they take flight only when they have seen the source of the noise and the sight is equally frightening. The visual impression is obviously more important than the audible one, so they are rarely disturbed by the most alarming noises or music from loudspeakers as long as the origin of the noise does not look frightening, though the same music coming from a visible brass band may easily put them in a state of mind preliminary to panic. However, like small children, horses appear to enjoy noise and, when they have to spend too much time in their boxes, they prefer noisy games in order to hide their frustration.

Horses are not frightened of any noises made by people which are similar to those they make themselves such as coughing or snorting, but they are easily frightened by unfamiliar sounds, even very quiet ones. One never whispers or pretends not to be around when horses are dozing or quietly feeding – they will never take flight if spoken to clearly. Since sights are obviously much more likely to make a horse take flight than sounds, the training of horses to traffic has to begin by getting them accustomed to the sight of cars coming towards them and the usually unaccustomed smell of petrol. Even when quite unpleasant, the stimulus of smell causes them to become alert, to snort or to *flehm* but, in order to flee, they seem to need a visual signal. The smell of blood is an exception, although some horses apparently do get used to it. One is liable to think that runaway horses or, in other words, animals in full flight are always well-bred horses, since they truly do possess the urge to flee rather more than heavy horses with placid temperaments. However, heavy horses may also shy and then suddenly erupt in a fit of uncontrollable panic, which does not manifest itself in quite the same way as the terrified flight of the lighter-bred horses, but the motivation is just as powerful. This may be due to the difference in the original habitat of these slower horses, since they are apt to jump aside to evade danger rather than to run away in fright. Neither the heavy horses nor the heavy-weight hunters show less sensitivity than their better-bred counterparts and, contrary to expectation, when one gets to know them better, they are almost more nervous and shy. If one does not recognize their fear as easily as one does in some hysterical, highly-bred horses then it is due to the superficial manner of observation common to many people.

# THE HORSE'S FORMS OF EXPRESSION

# 8

# VOCAL EXPRESSION

Before we study the miming expressions and gestures of the horse, and the individual 'faces' and different ways he carries himself, by which feelings and intentions are conveyed from one animal to another, we shall examine the vocal expressions of the horse — his actual language which, although inferior to ours, like that of all animals, consists only of a scale of different sounds. Principally, however, the horse's language has the same roots as human language, although it is much less important socially. The formation of a proper language requires a highly-developed intelligence, which even animals that are considerably more clever than horses do not have, for example, the anthropoid apes which, even when talking to each other, are only able to use a variety of noises.

## CAUSES OF VOCAL EXPRESSION

For us language is the most important medium of our existence and of our social contacts, including the babbling of babies and the inarticulate noises made by imbeciles. As far as horses are concerned, vocal expression is secondary to visual forms of expression and serves only as a possible basis of understanding when the visual signals are not clear, perhaps because of distance or if the animals are hidden from one another. The frequency of neighing is, therefore, dependent upon how much horses are able to see of their surroundings, which means that calling is either necessary or not necessary to keep members of a family together if they have been lost from sight. The frequency of neighing even depends upon the species to which the animals belong; some appear to be more 'talkative' than others and in this the donkey is foremost, whilst the onager is rather silent. To what extent the environment contributes to this will be seen when current studies of wild Equidae are more advanced.

## VOCAL CONTACT BETWEEN DAM AND FOAL

There are various contradictory opinions held about domestic horses: some writers think they are fairly 'talkative', others describe them as silent. My own studies show that their audibility depends very much on the time of day, how the animals are kept and, indeed, the age and individual peculiarities of the animals. Before they can even stand, new-born foals will call in a remarkably deep, soft voice and, after a few hours, can command a complete scale of sounds, since their audible contact is of greater importance to them than it is to older foals or fully-grown horses. We do not know how much the sounds which the mare utters, when she responds with soft, almost tender, low whinnies to the calls of her suckling foal, help to form its character. If small foals lose their dams, their calls are so loud and shrill that they turn into cries of fear. This shrill, spine-chilling distress call for the mare is heard in the frightened neigh of growing foals and even in fully-grown horses, and is obviously different to other calls. The 'boss call' of wildly rearing stallions, which we frequently hear in films featuring horses, are generally nothing more than a superimposed neigh of fear or possibly a terrified call and are in no way the proud challenging order of a herd stallion. I do not know exactly how the American producers do this, but I imagine that they take a horse away from the herd, as it is well known that isolated animals usually call loudly for help.

Vocal contact is, therefore, primarily heard between mother and offspring when the foal thinks it is alone. After a few weeks mares that have produced several foals are not so dutiful in replying to the foal's calls, and either ignore them or react very slowly. This conduct is part of the foal's education and, as he grows older, it makes the youngster more independent. The mare's reply to older offspring is conspicuously different to that given in the early hours of the foal's life and already sounds like the normal neigh of contact by individual members of the family, which the half grown or fully grown are wont to give.

## THE VARIETY OF VOCAL SOUNDS

Horses of all ages recognize each other from afar not only by the visual picture but also, very clearly, from the audible tone, strength and number of overtones, as well as the length and the whole tone-range, which an attentive human ear learns to differentiate in a fairly short time. The horse's neigh develops by degrees; one can almost describe it as a voice breaking during puberty, as yearlings still use the voice of their early days. During their second year they begin to use the lower tones of maturity, until as two- or three-year-olds they have generally attained the final depth of tone. At the same time, a vocal sexual dimorphism appears, so that stallions and mares possess different sounding voices.

The challenging neigh of the stallion is quite unmistakable, since it is clearly differentiated from that of the mare by its piercing, ringing tone. Besides this clear call, which is heard if the mare is out of sight or he hopes to see her at any moment, a stallion also uses a series of angry sounds, which seem to come from deep inside him and turn to a form of snorting or grunting as soon as he approaches a mare and begins to show off.

A mare in-season squeaks quietly in a defensive tone if the stallion becomes too rough and, if she is not ready, it swells to a much louder and longer squeal, and can develop into a positive yell of anger. Quarrelling mares use an angry trumpeting 'war cry' whilst they kick each other with their hind feet. Stallions, too, have a similar 'war cry' with rather roaring undertones and, if for example they have been hurt or bitten, they utter cries of pain or fear, when the inferior stallion takes to flight. On other occasions the normal neigh is usually one of disgust, as when horses out in the paddock on their own run up and down the fence calling, or it may be an impatient neigh at feeding time. These neighs range from a cheerful whinny of expectation to a fairly long, ringing, challenging roar.

The vocal tones of Equidae can scarcely be expressed in words, as anyone knows who has to do with horses. Of course, the different ranges of tone can be scientifically measured on an oscillogram, but for the ordinary horseman this seems to be far more abstract than a verbal description, since a 'language' is actually better understood acoustically. Besides the different neighs and single squeals, all Equidae possess a common snort of alarm, which escapes from them at the appearance of a possible or actual danger and is a warning signal to any horses nearby. We do not need to go into the details about the sighs and groans of great exertion, when speed is exacted or when covering a mare, or, above all, the groans of pain of a mare in labour because these sounds are common (although varying in intensity) to both horses and humans, as are those of sneezing, coughing and other bodily sounds. Human laughter seems to be an exception; horses cannot laugh and most of them react to a laugh with astonishment. I bought a doll for my small niece, which had a long realistic child's laugh, and we tried it out in the stables. To our great pleasure, it caused a number of the occupants to reply with a chorus of friendly and frequently repeated whinnies.

## AUDIBLE COMMUNICATION WITH OTHER ANIMALS

It is well known not only that Equidae will reply to the calls of their own species but that a horse, for example, will react to the bray of a donkey or zebra. Experiments were carried out by the Berlin Zoo with the help of a cassette recorder and it was found that horses responded loudly to calls made by their own kind, not so loudly to those of other members of the Equidae family – the more distant the relationship, the less interest was shown in replying. They became attentive, tossed their heads and pricked their ears at the neighs of distant horses but, when a cow mooed, there was little, if any, reaction. Only by constant close contact can non-related species come to understand one another, as we can see between humans and horses. We have already dealt with the one-sidedness of making the horse understand *our* language. We know he will soon recognize the voice of his rider or groom and the meaning of various words. We need not detail the ways in which quiet, soothing words can inspire confidence and the shattering effect of short, sharp orders.

## THE FACIAL EXPRESSION WHEN NEIGHING

Horses scarcely open their mouths as long as the tone is low and it is only when the voice is raised at length that they pull their lips back in the region of the upper jaw diastema – the gap or bar between the incisors and molar teeth – so that, when they neigh loudly, their mouths are wide open and the corners of the lips form a semi-circle. At the same time, the tongue is curved and the upper incisors remain covered by the long, mobile, upper lips, and the bottom teeth are free, as long as it is a non-aggressive neigh of contact. The molar teeth are not visible when a horse is neighing to show off. The neigh is always directed towards the object – no matter whether it is a member of his own species or a bowl of oats – the ears are always pricked, the nostrils wide open, the neck generally held high, and, as the tone increases, the head is held horizontally to strengthen the volume from the throat. For this reason singers, too, lift their heads, when high or loud notes are produced.

# 9

# FACIAL EXPRESSION AND PHYSICAL BEARING

## PARTICULAR ELEMENTS OF EXPRESSION

As we have seen, horses have a whole range of vocal expressions at their disposal, although these sounds are by no means equal to the importance of their comprehensive 'facial language'. This ranges from relatively plain, very obvious signals to very slight changes of the face that are scarcely noticed by humans. They cannot knit their brows as we do, and the area between the bridge of the nose, eyes and ears is immobile, but, in spite of this the play of the features – the ears, nostrils and mouth – is full of meaning for those standing near enough.

### *Play of the ears*

Different moods are best recognized by the positions of a horse's ears, signals that play an important part in the language between animals. Although the ears of the various Equidae, including domestic horses, vary in shape and size – from the tiny pricked ears of some pony breeds to the long ears of the ass and the spoon-shaped ears of the Grévy zebra – all signals of intention and mood given by the ears are understood equally well by all the Equidae species. Thus, when a horse lays his ears back as a threat, it is understood by zebras or donkeys as a warning. If we put our hands on our cheeks and allow three fingers to show above our head, we can imitate ear signals to which they will react attentively and their mood can change, if for only a few moments, because of the movement backwards or forwards of our artificial ears. Not only the human lack of ear signals, which may possibly appear to a horse to be a constant threat, but also the different ways a dog uses its ears can be misinterpreted, for example, when a dog lets its ears go back and its face breaks into a friendly grin, showing all its teeth, horses do not necessarily interpret this as a happy greeting or a challenge to play and it is usually regarded as a terrible menace.

The ears can denote and emphasize a whole series of moods, from the forward pricked position, denoting vigilance or curiosity, to the various positions backwards, until, when really angry, they lie flat and almost invisible on the horse's neck.

*Illustrations 104-117*

Plate 41  Threatening expression and carriage.
*above left:* The Welsh cob stallion Boschveld Aladin steals along with wicked intentions; the ears are pricked inquisitively, nostrils and mouth pursed threateningly.
*above right:* Holstein mare Aischa drives off her Trotter enemy in a first-class traversale.
*below left:* Ponies with concave profiles, round eyes and short ears never look as dangerous when they threaten as
*below right:* do long-faced, long-eared, larger horses wearing the same expression, especially when they show the white of the eye.

Plate 42  Very threatening expression.
*above left:* The Holstein mare Ganda's ears have vanished, even when seen from the front.
*above right:* The Trotter mare Ina produces a grim expression with her wickedly pursed nostrils and mouth.
*below:* The threat is carried out – Düne, a Hanoverian mare, bites.

(Continued page 167)

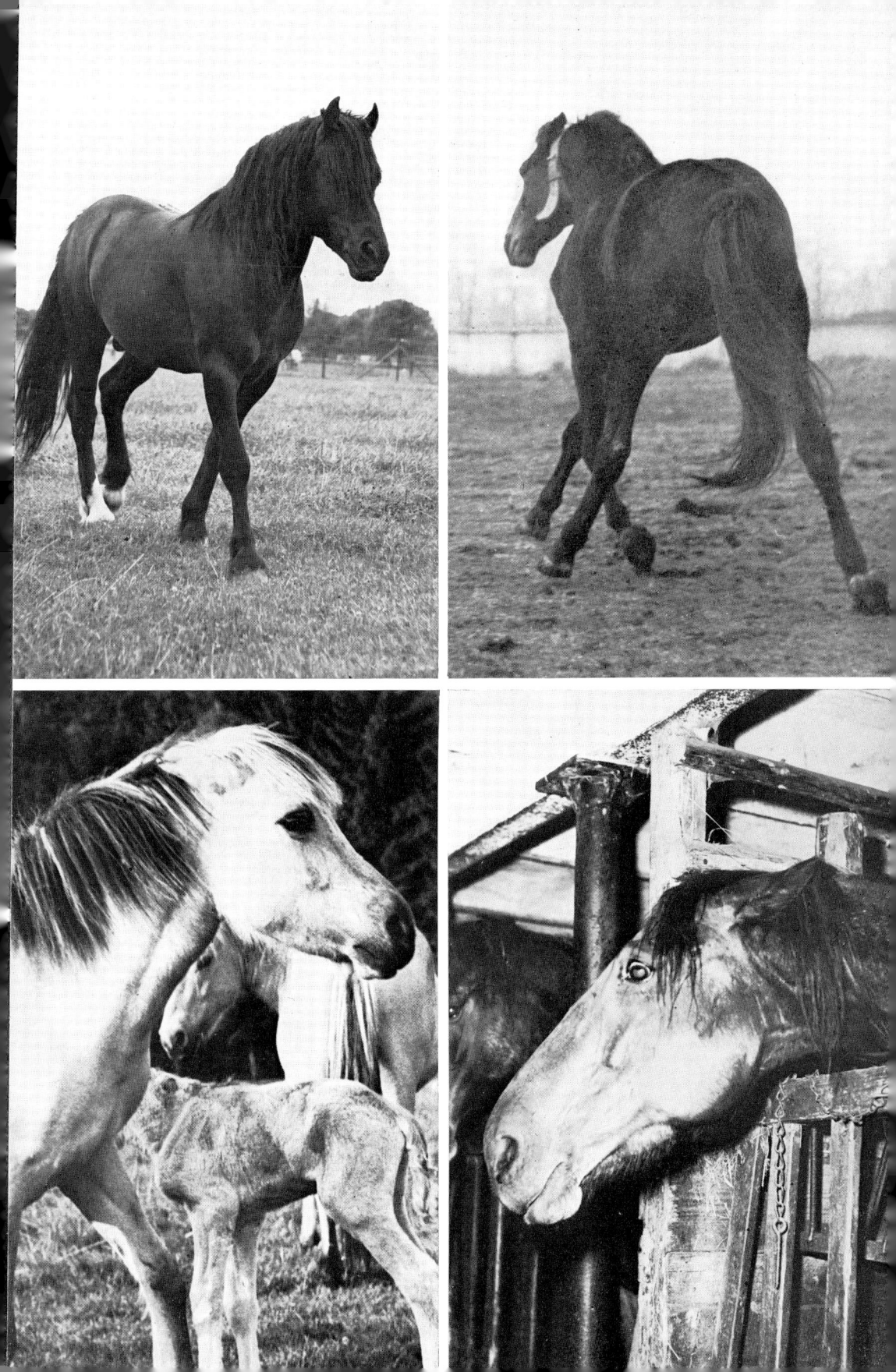

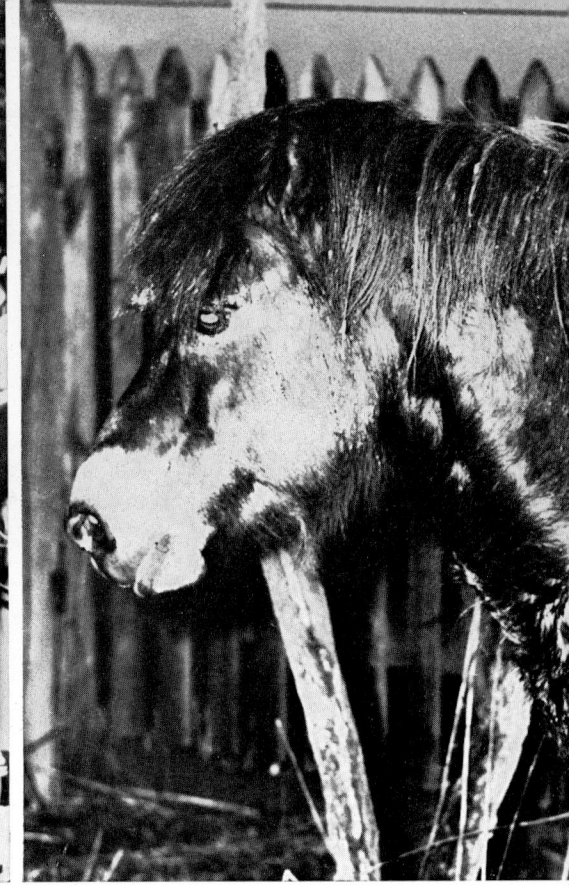
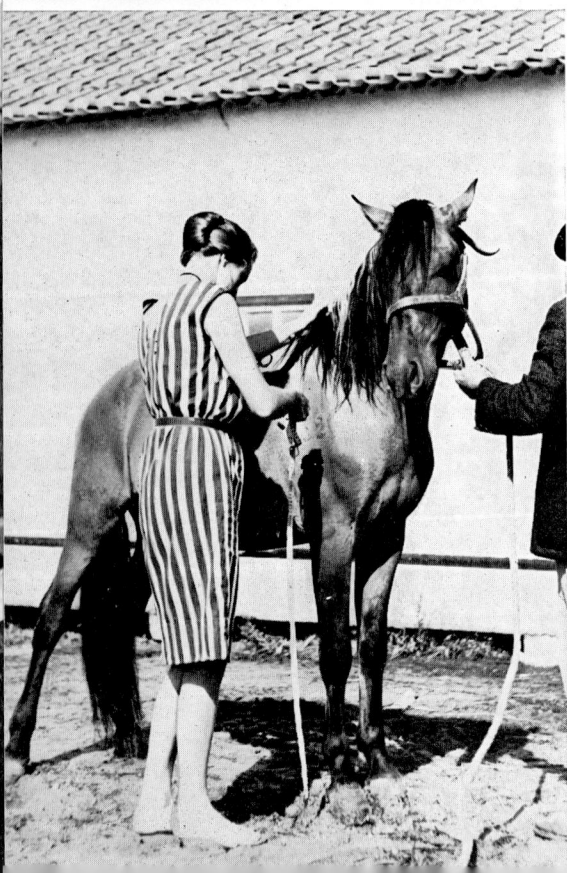

Apart from these different positions of the ears (moving in an arc, like a swinging barometer needle, going forwards and backwards), particular moods such as inferiority, tiredness or concentration not adjusted to an audible stimulus are often shown by the ears drooping sideways in varying degrees from vertical to horizontal. To these combinations of direction there is in addition the subtle moment of mutual understanding, when the opening of the ear is turned towards or away from the object. Depending on certain exterior stimuli or an automatic response, the ears turn more to the side or droop downwards.

Plate 43   Young horses show submission by their bearing.
*top:* A slight threat by the herd stallion Endo is sufficient to put Asko in his place. The two-year-old reacts meekly, wearing an expression of submission.
*centre:* Three Welsh yearling colts humbly bow their heads as they pass the stallion. Their submission is characterized by their lengthened necks, ears pointing sideways and backwards, and exposed incisors as they chew rapidly. The fact that they try to make themselves look smaller by drawing in the hindquarters denotes their subservience.
*bottom:* The Trotter foal Intrigant is bitten in spite of his submissive chewing. His curved tongue can be clearly seen. His inquisitively pricked ears show that he is not completely submissive.

Plate 44   Submissive and frightened faces.
*above left:* The Andalusian parade stallion, belonging to the famous Angel Peralta, in spite of his proud bearing shows submission to his rider by the position of his ears.
*below left:* The Sorraia stallion Navigante with submissive ears and tense, easily frightened bearing, ready to flee as he is measured at the withers.
*above right:* The face of an Exmoor stallion Carabineer who has been free, is now cornered and is terrified. (Photo. E. Trümler).
*below right:* The semi-wild ponies from northern Spain become very frightened when they are driven into a corral each year.

### *Twitching of nostrils and muzzle*

Besides the visible play of the ears, the twitching of the nostrils and muzzle is equally a form of expression. Although this is not so readily understood by humans, the slight nuances are both seen and interpreted by the horses themselves. The different degrees of expansion of the nostrils, the drawing back or curving forwards of the outer corners of the nose, the extension or retraction of the lips which is done passively or as an element of an active alteration of the lines of the mouth, the varying shapes of the corners of the mouth and the openness of the mouth, possibly showing the upper and/or the lower incisor teeth – these are all part of the inaudible language of horses when they are very close together.

### *Meaningful eye signals*

Compared to other facial features, the eye holds far less meaning, since the value of its signals at a distance is nil. Apart from passively blinking, opening and closing the eyelid and the extent to which the eyeball is retracted in its socket, the only direct possibility of expression is by rolling the eyes as a means of intimate understanding. There remain only the vague and emotional, from the kind and maternal to the wicked and mean, expressions in the eyes, which at least give humans some idea of the character, temperament and mood of the animals and their physical well-being or illness. Otherwise the eyes are never used for any kind of mutual understanding since, as we said before, the horse's bearing and other expressions are sufficient for the horses themselves, as well as for people who have learnt their language. The reader can see this for himself from the illustrations in this book, if he covers the horse's eye. None the less a great deal of nonsense has been written about this particular organ. Some horse-lovers are quite unable to free themselves of ideas which they developed when young and they usually see only the reflection of an inner virtue in the 'dark, velvety, liquid eye' directed kindly in the direction of the human concerned! It is obvious that we, whose interest in the horse is just as great but rather more objective, should not attempt to humanize him by seeing all his virtues in his eyes, but should see him and all animals as harmonious entities, and we must look much more closely in order to interpret the expression of a large or small, a round or elliptic, equine eye.

### *Changes of bearing and silhouette*

Horses react immediately to changes of the normal silhouette, as we know already from the fact that raising the tail is a signal to gallop and the wide-legged stance of the mare shows her willingness to mate. In this category, besides the different ways of raising, lifting or squeezing down the tail, there are various ways of carrying the neck, i.e. bending, stretching or lowering, which show aggressive or defensive moods. The position of the head is easily recognizable against a background from afar, and indicates the object of the horse's interest and his frame of mind. To what extent the outline suffices or whether movement of the contour is necessary to cause others to react depends upon the situation at the time. Horses also have to become accustomed to the sight of a rider, because they cannot at first differentiate a horse from a human.

## ORIENTATION

There are many different forms of expression that characterize the enormous complexity of orientation behaviour. Equidae are not just content, when grazing, to inform themselves about things concerned with eating and drinking; they try to satisfy their curiosity with all their senses and to safeguard themselves as far as possible against the unknown. Assessment over a distance, which is usually directed to moving objects, relies chiefly upon optical impressions, assisted by the sense of hearing as the interesting object comes closer, at which point the sense of smell comes into action as the third source of information in addition to seeing and hearing, whilst taste and sense of touch are of less importance.

### Assessment by sight and sound

Hearing plays an important rôle in orientation, whether the object concerned is near or far away, and the attentive play of the horse's ears, even for an inexperienced observer, is very instructive. To establish the direction of a familiar noise it is usually sufficient to turn the ears towards the source, without having to use the eyes or the sense of smell. Horses demonstrate this most clearly out in the paddock, when they follow the progress of the approaching groom, although apparently they continue to graze contentedly. This behaviour will change according to interest or uneasiness, i.e. how near the object approaches, and the horses stop grazing, jerk their head upwards, prick their ears and look directly at the disturbing object. With this typically attentive movement, the nostrils are partly open to take in the scent and, from the position of the head, the expert as well as the other horses can easily determine exactly where the enemy was spotted. The characteristic position of attention is not the only clear indication to the other members of the herd to be on their guard but the clear silhouette of the horse's body also shows the direction and distance, though the exact position of the object is always a little further away than the horse's nose appears to indicate. The further the horse gazes into the distance, the higher the head is held; the nearer the source of disturbance, the lower the head will be bent. It remains to be discovered, as far as the different Equidae are concerned, how far the shape of the skull, the position of the nostrils and the eyes, and their ability to focus, are decisive factors in the way the animal throws up its head. A larger skull with the eye in a higher position, like those of some heavy horses and wild Equidae, allows the head to be carried lower, whilst the smaller Arabian head, with its proportionally broader cranium, like that of the foal whose profile is somewhat convex above the eyes, results in the head being held higher and more horizontally.

The horse's face also expresses with varying intensity his degree of interest in the surroundings. As long as events are harmless there is almost no expression; participation is shown by a change of head and neck position, and by the position of the ears. In real or supposed danger the eye-lids are opened wider than usual, the nostrils more inflated and, above all, the general alertness of the horse gives the impression that he is really wide awake. Other Equidae are just as alert, but this fact is not so noticeable for anatomical reasons, and I do not think it is always correct to assume that such Equidae are less intelligent.

If the origin of a disturbance, a strange noise, sight, taste or smell cannot be explained clearly, horses wear a definitely puzzled expression, distinctly evident from the nervous movement of the ears from side to side, eventually turning in different positions. The play of the nostrils and the uneasy glance portray the great inner conflict of the horse – divided between irritation, indecision, aggression and fear. Very young foals often look like this since on their first excursions they have to cope with a great number of new and foreign impressions.

## Assessment by smell

Horses have a wider range of vision than humans because their eyes are placed at the sides of their head, although immediately in front of the nose there is a small area where they see nothing at all. If they want to examine something lying very near them, they have to decide either to examine it optically or to use their sense of smell. Depending upon whether the object is recognized or is strange, horses approach with confidence or more carefully with the head lowered, nostrils opened, ears pricked and eyes wide open. Stallions, which are very exact when examining the dung or urine places left by their comrades, hold their nose immediately over the rising scent, whilst the ear-openings are turned sideways, so that, though concentrating on the sense of smell, they can still control possible danger beyond their range of vision. Their outline is different to that when grazing and, when they lift their heads to *flehm* – as they always do when smelling these dunging places – the particular position of the head and neck seems to me to be the signal for all the other male members of the herd to become interested too.

One rarely sees horses trying to catch a distant scent; if they do smell something carried on the wind, they toss their heads high, sniffing with wide nostrils. It is clear that in this case the nose plays a larger rôle in orientation than the eyes and ears. One of the most impressive gestures of horses of all age groups and both sexes is the so-called *flehm*. When horses *flehm* after smelling something interesting or strange to them, they turn their top lip so far upwards that their nostrils are closed completely. If at the same time the bottom lip hangs down, the closed incisors can be seen in all their glory. They gradually lift their head upwards, often until it is almost vertical, move it to left and right, and shut or roll their eyes in full enjoyment, whilst their ears turn in all directions. The reason for this remarkable action, when the air is held and the nose blocked, is the position of the Jacobson organ at the base of the nose. This organ is a mucous-covered tube composed of cartilage and filled with a vital liquid. As far as I am aware it is not known what special odour it is which causes such an intensive olfactory examination. Stallions always *flehm* after smelling the sexual organs of mares which are in season; mares, on the other hand, *flehm* considerably less.

### Assessment by taste and touch

As soon as their interest has been awoken, Equidae seek to inform themselves with all their senses. Foals, especially, possess a marked curiosity and always investigate strange objects with their muzzles, nostrils and tongues, whilst the long hairs on the muzzle and chin, which have nerve cells at the end, help too. Some foals and even fully-grown horses examine objects with their forefeet. Perhaps they hope by this means to 'loosen' the object in order to be able to smell it better. It can also be a gesture of impatience. Horses have an excellent sense of taste and with their tongues can separate anything foreign in their food, for instance, when grazing, an unwanted weed or stone is carefully allowed to drop out of the mouth. Even small weed heads, tiny pieces of paper, straw, etc., are carefully sorted from the oats which they are in the process of eating.

## FACIAL EXPRESSION OF FATIGUE

Boredom or tiredness are shown by yawning in varying degrees – just as humans do. The mouth is opened to its maximum with the upper lip drawn back over the gum; at the same time horses usually shut their eyes and push their lower jaw sideways once or twice. Both humans and horses yawn in the same way and for the same reasons – and a horse will very soon copy a yawning person.

When dozing in the position already described, with the croup sunken and one or other hindleg resting, the horse's face and his muscles are relaxed so that even young, strong animals as well as very old mares often allow their lower lip to hang. The nostrils are half-closed, because the breathing is lighter, the eyes almost covered by the lids and the ears stand upright – when completely relaxed the ears hang sideways with the opening downwards. Also, when asleep lying down with legs underneath the belly, with the head either held free or resting on the mouth and front teeth, as well as in deep sleep with the legs fully stretched out, the difference in the facial expression from that of dozing is apparent from the fully closed eyes.

## FACIAL EXPRESSION OF GREETING

### *Greeting ceremonial*

When two strange horses meet, they greet each other according to established ritual, which is rarely carried out in its entirety. When the ceremonial is fully performed, they approach one another with ears pricked and this visual impression produces a reaction of curiosity and interest. If both then show a friendly attitude and avoid all signs of aggression, after mutual smelling of noses, the shoulders, flanks, dock and sexual organs are also smelt. In this second phase the sense of smell is of primary importance, although both horses are watchful for a possible change of mood in the other. This greeting develops into a mutual nibbling of the skin and ends the ceremonial, which is performed only once. When a newcomer enters a herd, the family stallion or the lead mare will begin the ceremonial; all the others follow according to their rank in the same way.

As a rule this greeting is broken off fairly soon by one or other displaying a desire to show off and impress the other by threatening gestures, and thus establish his superiority, so that the other is forced to submit and admit his inferiority. Interestingly enough, the greeting ceremony runs its course to the end if one partner is merely a model. As Grzmick's experiments with domestic horses and zebras in the Serengeti show, which Trummeler also confirmed later with zebras, Equidae will greet even simple two-dimensional pictures as proper members of their species, and this is how the importance to horses of seeing silhouettes was first discovered. However, the living horse naturally enough loses interest as soon as he finds that the artificial model has no smell and does not react as it is expected to.

Many experts think that horses wear a particular expression on their face when going through the greeting ceremonial, but their moods change so rapidly from one gesture to another – showing off, aggression, submission, curiosity – that no single expression applies from one horse to another. Thus written observations are often contradictory.

### Neighing to impress

The prehistoric horse possessed longer canine teeth (tushes) and, when it opened its mouth to greet another member of the species, it purposely showed these fighting teeth with which it delivered vicious bites. The modern horse possesses very much smaller and more inoffensive tushes, and, when the horse bites, he uses his incisors. When a horse neighs at nothing in particular, hoping to hear an answer however distant, he scarcely shows his teeth, but, should he neigh a greeting to another horse not yet knowing its rank, he will open his mouth much wider to show his insignificant canines. The action is the relic of an ancient instinct when these teeth were used to bite and served as a warning that the owner was alert. The tapir, the nearest relative of the horse, yet one of the more primitive animals, still uses this archaic form of aggression and draws his mouth so far back that the canines are fully exposed. However, whilst the tapir is actually threatening aggression, the horse with his ears pricked forwards moderates the aggressiveness of the greeting.

Even the Arabian horse shows the same archaic gesture, although the diastema may be longer and the teeth not quite so visible. All stallions of all breeds carry their head in a special way when they neigh to impress, even if they have a bit in their mouths: the neck is held high with the head held perpendicularly; whilst for the contact neigh, the head is still high but is held horizontally. Experts confirm that zebras have a similar facial expression when greeting each other and, when this is very intense, the mouth is opened still wider.

## MANNERISMS TO IMPRESS

When horses want to show off, they curve their neck in order to appear more powerful and bring their head down to their chest. The ears are pointed towards the object that has to be impressed, the nostrils are partly opened and the tail is carried well. Equidae impress or show off only at the trot and I believe that the trot was developed for this purpose. With the exception of the Trotter, which has been selectively bred for this particular gait, free-running horses usually move at the walk or canter. Of course, such free horses do break into a trot before setting off at the canter or gallop, but it is only a transitionary movement to help them obtain sufficient impetus and has nothing to do with showing off. The high showy trot, which even very young foals can do, although normally they can canter from a stand, extends from dancing on the line whilst being led in hand – the favourite gait of stallions in front of a mare –

to the piaffe, a high-stepping trot on the spot, the cadence and the action depending upon the type of horse.

When horses are prevented from moving forward by some kind of obstacle, they are often obliged to do the piaffe in order to impress their contemporaries. The half pass, a traversale, is also a very impressive movement and horses will freely execute this movement at the trot without a rider. Mares show off to the same degree as stallions, but I think they are less inclined to do the piaffe. Perhaps their behaviour derives from the masculine repertoire, since autocratic high-ranking mares sometimes show off with the same effective behaviour mannerisms as stallions do.

If horses want to reach their objective quickly, they may forget to show off and the trot turns into a strong or 'butcher boy' trot. This gait is not the high action of the passage, but is a long, stretched action of all four legs and feet, and, since the horse is not collected, he usually pokes his nose out and this 'impressive' action seems to be common to most wild asses and members of the *equus hemionus* species.

Earlier in this text I wrote of making an exception of Trotters since they are selectively bred to trot, but that they too have inherited the natural tendency to show off at the trot can be seen by the following incident in my own stable. One of my Trotter colt foals had to manage without a playmate for the first year of his life, because of which he always walked or cantered, although his parents were extremely good Trotters and his mother rarely cantered in the paddock. We decided that he had been alone too long and turned out another colt of the same age, and it was then that my belief that the trot was originally a movement designed purely to impress was again absolutely confirmed. The home-bred yearling immediately tried to impress his rival by arching his neck and, fully collected, he broke into a high cadenced trot. It is interesting to note that he retained this gait from then on as an alternative to the walk and canter.

Snaking the head is another movement aimed to impress and indicates resentment or animosity. It is a different movement to the one used to shake off flies, when the nose is perpendicular. When 'snaking', horses are in various stages of collection turning slowly in circles and figures of eight. Such action does not necessarily indicate that the horse will attack, but it does mean that he has surplus aggression and it is, therefore, generally noticed in vicious animals of either sex which are separated from their companions and want to reach them or which do not like the people in their vicinity.

## ATTITUDES OF AGGRESSION AND SUBMISSION

### *Threatening attitude*

There are few attitudes shown by the horse which are as easily recognized, even by people, as the attitude of aggression in all its forms. Laying the ears back is the most usual and, although the animal may not actually bite, it shows that his mood is one of resentment and is either offensive or defensive. If the horse is in a really bad humour, the ears are laid so far back that they cannot be seen in profile at all and this is a gesture that is rarely overlooked by their companions. We find that this attitude is the most dangerous when displayed by horses with common heads and long faces like those of some warm-blooded and many heavy horses, but it is also seen in wild Equidae, which frequently become aggressive, whilst the same gesture has far less effect on us when adopted by ponies or by Arabs with their delicate concave profiles. Added to this, pony ears are so short that their value as a signal of mood is less impressive than those of the extremely long-eared wild ass.

Besides the very noticeable laying back of the ears, the expression of the nostrils and mouth is also important. In this mood the horse draws the corners of his nostrils back so impressively that this movement alone appears as a serious threat. The second phase is shown when with closed lips the corners of the mouth are drawn downwards, and the expression becomes increasingly threatening when the horse opens his mouth, raises his lips and exposes his incisor teeth. If the horse feels inquisitive as well, the ears may even be pricked, so that only the nostrils and mouth show his real intentions, which are recognizable to an expert.

Some horses become extremely aggressive when we bring them food or water, when, as we think, we are doing them a kindness. This animosity, which is a form of anticipatory jealousy, reminds us that Equidae, in spite of generations of being fed by humans at set hours, have not become accustomed to the loss of the natural habit of feeding throughout the day. Their apparently ungrateful behaviour shows once again how much they are still influenced by their primitive instincts, although they are domesticated, and how difficult physical adaptation is for animals of limited intelligence.

The way in which a horse moves can be as threatening as his facial expression. This threat is reinforced by lifting a foreleg or a hindleg and swishing the tail before the animal actually strikes or bites. He then gallops at his opponent and carries out his intention. Stabled horses have to waive these preliminaries because of lack of space and, therefore, their lightning bite or kick can easily catch unawares an inexperienced person who has not noticed the signs.

A special form of threat is the posture used only towards mares by the stallion, which at a walk or other pace will make his mares change direction or bring them to heel. The obvious silhouette of the stallion in this pose, with head and neck stretched below the horizontal, almost touching the ground, ears laid back, the head snaking from side to side, and tail held high, is recognized at once by the mares and obeyed immediately.

## Submissive attitude

All Equidae have developed a special posture to show their submission and it varies according to the age of the horse. Depending on physical and psychological maturity, foals up to approximately two years of age show the so-called submissive face. They make chewing movements with the lips, whilst showing their teeth, and although their teeth do not actually chatter, as ours would, a suckling noise can sometimes be heard, as the tongue is visibly drawn across the roof of the mouth. The nostrils are drawn open and the ears laid back or sideways. Added to this facial expresion, there is the expressive position of the body. If youngsters are threatened, they jerk their heads up before they begin their chewing movements, and out of fright they squeeze their tail between their legs, pull their hindquarters in and almost allow all four legs to collapse, so that the target is smaller for the attacker to bite.

Unlike other young animals, foals show their submission in a standing position, probably so that they cannot be driven away from the herd. Their childish make-believe behaviour helps to ward off an attack by an older animal. It still has to be discovered why older animals should make such fierce attacks or persecute young animals, but it is possible that domestication has had a damaging psychological effect on the former. Foals submit at once with the right gesture of chewing, but older youngsters really have to be treated firmly until they show a proper submissiveness. It does sometimes happen on such occasions that these youngsters show the expression of submission on their faces. When a barrier lies between themselves and the older horse, they also wear a form of primitive 'menacing' expression, but without actually showing their teeth, whilst their ears are pricked in a friendly way. These two instinctive expressions are contradictory and are due to inexperience, since the youngster may be uncertain what is expected of him.

Fully-grown horses show their submissiveness towards a superior by squeezing the tail between the legs – in fact, they wear a 'hang-dog' look, especially in the position of the ears, which are more or less horizontal, depending on the degree of submission required. This submissive expression is seen in the relationship between humans and those horses which are predestined to be inferior, or whose own independence has been so destroyed that they are no longer able to think of resistance. It is significant that the Spanish *vaqueros* do not speak of trained or untrained horses but of horses that have been 'broken' or 'not broken'. The feeling of panic is related in a very exaggerated sense to that of submission. Horses rarely feel true panic akin to the fear of death amongst themselves, because the aggressor reacts to the gesture of submission. Genuine panic is caused by animals of other species which do not recognize or which ignore submissiveness and, above all, when the horse is cornered by man and there is no escape. When wild Equidae are cornered and caught, the same form of panic arises as that of a domestic horse which has never seen a halter and is tied up and left; in the latter case a broken neck could easily result. We can only guess at the extent of a horse's fear, but we can see it reflected in his staring eyes, his quivering nostrils and forced breathing, and the position of his ears also shows that the animal is thoroughly confused and disorientated. The ears are not directed towards the source of danger but sideways and backwards. Hearing appears to play no part when a horse is very frightened, as it is only what he sees or smells that is of real importance, although a trained horse that is fearful will respond to a soothing voice and hand.

## EXPRESSIONS OF AVERSION AND LETHARGY

### *The surly face*

One often sees a horse wearing a surly and bored expression on his face. A range of moods can be seen in this expression from plain bad temper and boredom through of lack of work to plain dozing. The main purpose of this expression is to indicate that the horse wishes to be left in peace and not to be bothered.

### *Head-shaking against flies*

On a hot day flies can be a plague to horses, as they crawl around the mouth, ears and especially the eyes. The animal wears an expression of irritation and resignation, often with the lower lip hanging down showing the lower incisors, shak-

*Illustrations 118-133*

Plate 45  Bored faces.
*above left:* Warm-blooded mare plagued by flies.
*above right:* The bad-tempered face of a Fjord pony mare.
*below left:* Exhausted face of a Trotter gelding immediately after a race.
*below right:* The exhausted face of a Thoroughbred stallion after covering a mare.

Plate 46  Expressions of pain.
*above left:* Fjord stallion Findo is very lame behind.
*above right:* Holstein mare Aischa with slight colic.
*below left:* Haflinger mare Heidi is very tired after giving birth – she is trying to cleanse.
*below right:* Warm-blooded mare Gobi during the climax of labour pains.

(Continued page 183)

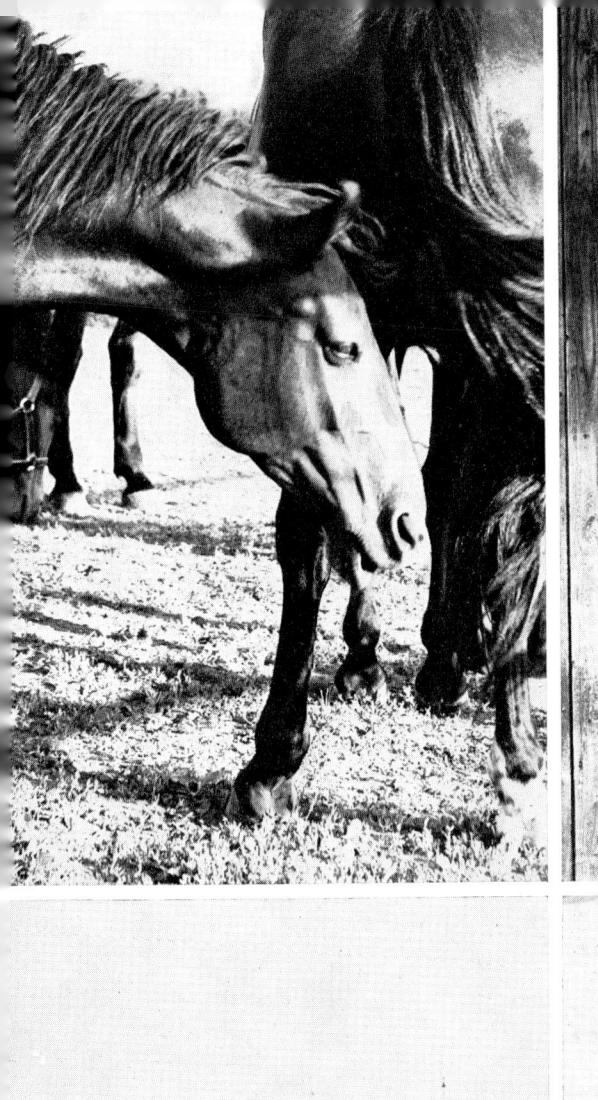

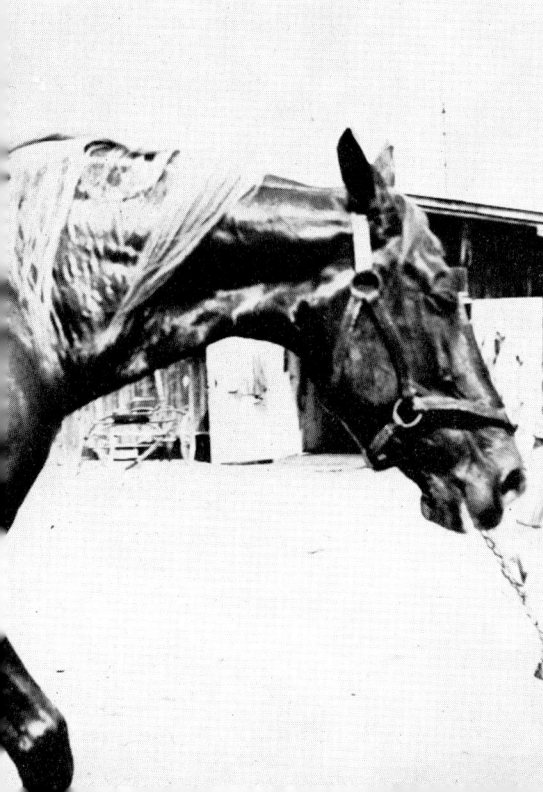
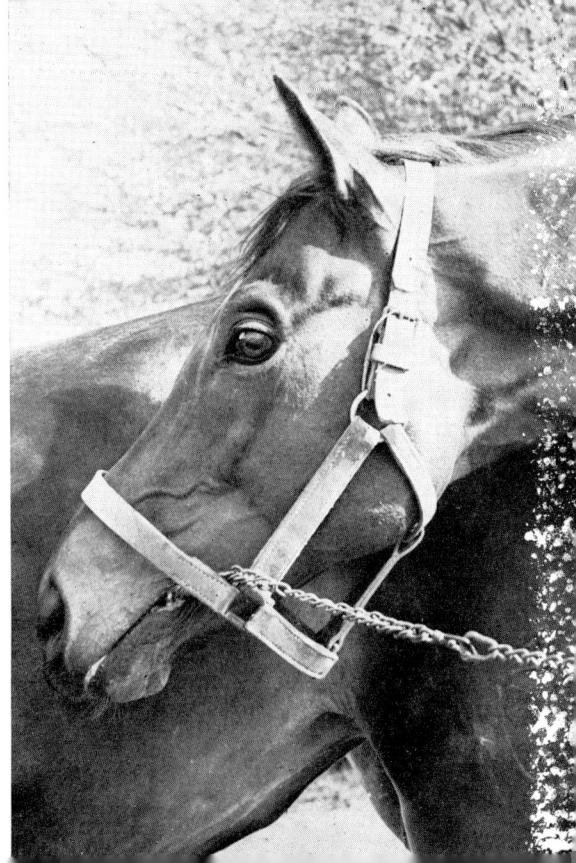

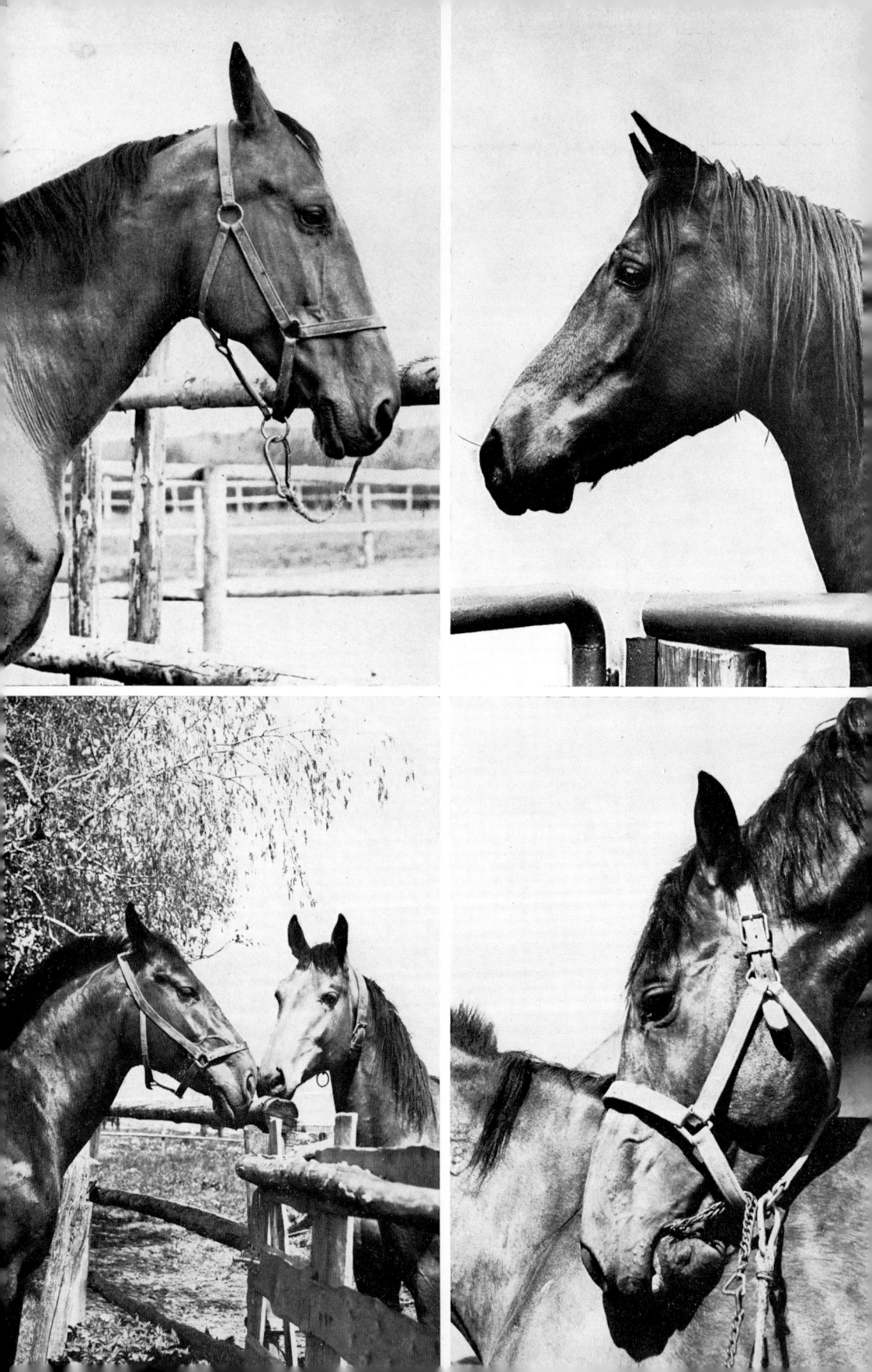

ing his head and occasionally rhythmically flapping his ears to try to remove the flies, knowing full well that it is in vain.

## The exhausted face

After great physical and psychological exertion, usually when one has demanded more than they are capable of, horses often wear an expression of complete and utter weariness: the nostrils are wide open, showing the pink skin lining and dripping with sweat, the eyeballs are sunken and the eyes dull or half-closed. The head hangs down and the ears hang sideways; the body sways and the gait is stumbling. It is obvious that they have given everything in response to the demands made of them. Trotters and Thoroughbred horses are sometimes extremely tired after an exhausting race and it is then that this expression can be seen.

Plate 47  Expressions of pleasure.
*above left:* Playful expression on the face of the warm-blooded foal Sorbas shows her intention to nip.
*above right:* Trotter foal Intrigant rubs himself with voluptuously twitching upper lip.
*below left:* Playful face free from aggression of the three-year-old pure-bred Arab stallion Mahomed.
*below right:* As opposed to the solitary coat-rubbing of Intrigant (above), the Icelandic yearling pony Skugi shows his desire for social coat care by his exposed lower incisors. In this case, his partner is a human scratching his crest.

Plate 48  Mating expressions.
*above left:* Holstein mare Ganda wearing an unmistakable mating expression.
*above right:* Pure-bred stallion Hamin purses his upper lip expectantly.
*below left:* The Oldenburger mare Alda flirts with the two-year-old Holstein stallion Radscha. Her head, ears and neck show that she is ready to be mated.
*below right:* The 'sensual' expression is similar to the 'nibbling' expression.

## The expression of suffering

A tired expression can easily change to one of pain and a fact of considerable significance is that there is no play of the ears. According to the amount of pain and the length of an illness, horses' eyes become smaller and, in chronic cases, sink into the socket, whilst the eye itself is dull and has a distant, almost lost expression. The nostrils become smaller and as pain comes and goes, they open and close with the breathing, which is often short and sharp. As the teeth clench, the face and mouth muscles become tense and the lower lip hangs down. The ears are turned sideways and backwards, whilst the sick animal appears to be looking and listening to its inside pain, taking very little notice of its surroundings. When they have a colic, they paw the ground and throw themselves down, and the ears are carried fairly upright, slightly backwards with the opening to the side. Once on the ground, horses with colic will try to roll, and a twisted gut and death may result. For this reason, they must be kept on their feet and walked around. With acutely painful lameness, the ears are laid further back. It is not yet known how far other localized illnesses affect the expression of the face.

## The look of a mare in labour

When a mare is in severe labour, she will be lying on her side, possibly groaning and moving her legs in the air. Her tortured face adds to the sum of her labour pains. Every breath appears to be forced through tense nostrils and, as the pains increase, when the breath is held they become smaller. The lips are pressed together and the chin is so tense that it resembles a ball, whilst the upper lip appears to overlap both incisors and lower lip. If the labour pains are very strong, the upper lip rises as it does when the horse tries to *flehm* and the clenched teeth are then visible.

## EXPRESSIONS OF PLEASURE

### Mutual grooming face

The wish to indulge in skin or coat care with a partner, apart from the diagonal approach, is shown by an unmistakable cleaning or nibbling facial expression. The horse pushes the upper lip so far forward that it very much resembles that of the tapir, exposing the lower incisors, whilst great care is taken not to show the upper incisors, in case the signs of his intentions are misunderstood for those of aggression. If all his teeth were visible, it would mean, as we know, the highest degree of aggressiveness. When horses are rolling energetically, the upper lip is also sometimes stretched out, flapping to and fro in pleasure, but, since no one else is being encouraged to co-operate in this exercise, the lower incisors are usually covered. During solitary and social grooming the ears are held in a passive position.

### Playful or cheeky face

Horses which do not have to spend the entire day grazing in order to get enough food but are stable-fed or are still suckling foals are often high-spirited and ready to play. When colt foals have been engaged in a session of mutual coat nibbling, they will show their intention of starting a scuffle or chase by turning and standing parallel, heads together – instead of head-to-tail. At the same time, they wear an indescribably cheeky expression, as anyone who has dealt with very young horses will know: the nostrils become inflated to varying degrees, the upper lip may be in its normal position or stretched out ready to nip the other foal and the eyes are very wide and alert. The foal throws its head up, ears pricked, without arching the neck, because, of course, there is no intention of frightening the other. It is only when the game changes to aggression that the facial expression also changes.

## Mating face

Since feelings are closely connected to a horse's frame of mind and are independent of any form of expression, they are reflected in all types of faces – and since feelings thus expressed are very fluid, it is difficult to photograph them. All feelings of satisfaction are shown by male animals, for example, with the top lip pushed up. Stallions wear an expectant look just before seeing or covering the mare. This rather sensual expression brings the lower incisors into view, especially during the intense mutual nibbling, though in this case the eyes always remain wide open and do not become small slits, as with the ordinary nibbling face, although their appearance is fairly intense. The stallion clenches his teeth on the mare's crest or shoulders, at which point the upper incisors are visible. The stallion becomes very excited when actually mating and afterwards has a noticeably tired expression.

With the exception of horses themselves, all other Equidae mares have very solemn faces during mating, when, according to the species, they open their mouths very wide, so that upper and lower incisor teeth are visible. The difference between this facial expression and the one that threatens is that the ears are not laid back, but submissively hang more to the side, with the opening downwards. This remarkable expression, which I prefer to call the 'female mating face', because zebras and she-asses do not look like this throughout the period of being in-season but only when the stallion is about to cover them, is not seen so clearly in domestic horses. Their expression is limited to the posture adopted, as described in a previous chapter, the moment they see or hear a stallion neighing shrilly and this is a help to the stud groom, who can tell if the mare is ready. The stallion is also aware of the effect that the mare wants to create from the way she holds her head and ears, and can thus tell that she is ready to be mated. However, an experienced stallion will take nothing for granted; he is aware that her expression can rapidly change to fear, when she may suddenly kick. Some stallions will approach from the front with friendly gestures and short throaty noises and only when he is sure that she is willing does he actually mate. On recognizing a stalion which she already knows, it is quite common for a mare to turn in his direction and whinny a greeting, which will in turn be answered with a great show of pleasure by the stallion.

I have tried from my own observations during many years of close contact with horses in their natural form to make more comprehensible to the reader such knowledge as we have at present about the behaviour of Equidae. Although I have purposely omitted the vast area of differences between the horse and rider, I hope I may have helped to make the language of the horse clearer to the average person, who will perhaps also be encouraged to study this exceptionally interesting subject.